Measurements, instrumentation, and data transmission

A text for the OND in Technology (Engineering)

B. F. Gray
Dean of Engineering, The Hatfield Polytechnic

Longman
London and New York

Longman Group Limited London

*Associated companies, branches and representatives
throughout the world*

*Published in the United States of America
by Longman Inc., New York*

© Longman Group Limited 1977

First published 1977
Second impression 1979

Library of Congress Cataloging in Publication Data

Gray, Bernard Francis, 1924—
    Measurements, instrumentation, and data transmission.

    1. Mensuration. 2. Measuring instruments. 3. Data
transmission systems.   I. Title.
T50.G69   620'.004'4   76—49922
ISBN 0 582 41065 7
ISBN 0 582 41066 5 pbk.

Set in IBM Journal 10 on 11pt
and printed in Great Britain by
Richard Clay (The Chaucer Press) Ltd,
Bungay

Preface

This is the second book written specifically for a subject in the OND in Technology (Engineering). It follows very closely the syllabus of Measurements, Instrumentation and Data Transmission published by the Joint Committee responsible for the course.

Inevitably it has some overlapping features with other subjects in the OND scheme, logic circuits and electrical measurements being two examples.

The level of mathematics assumed is minimal although it is expected that students have at least met complex numbers and are aware of certain relationships in the field of probability and statistics.

Although the book is written for a well-defined market it is hoped that it will appeal as an introductory text at first-year level for many students reading for a degree in any branch of engineering. Technician students in certain City and Guilds courses involving Instrumentation and Control topics should find much of the text useful. The small number of students in the Higher National Diploma courses in Measurement and Control will also find the book of some use.

Once again it is necessary to thank colleagues who have made helpful comments on the book which have resulted in improvements in the way a number of topics have been presented. It is also necessary to thank the authorities in the colleges providing the OND (Tech.) for their permission in allowing certain examination questions to be published. The source of these questions has been noted. It is also necessary to acknowledge the source of some of the excellent mechanical engineering diagrams — mainly from R. L. Timings' books, *Mechanical Engineering* and *Basic Engineering*.

Finally I must once again thank Marion Dance who typed the manuscript so well.

B.F.G.
Harpenden, 1976

Acknowledgements

We are grateful to Thomas Nelson & Sons Ltd for permission to reproduce copyright material from *Electrical Engineering Principles* by B. F. Gray, published by Thomas Nelson & Sons Ltd, 1970, and Hodder & Stoughton Ltd for permission to include some figures which are based on diagrams from Engineering Measurements and Instrumentation by L. F. Adams.

Contents

Preface   iii

Acknowledgements   iv

1   General principles of measurement   1
Standards and fundamental units. SI units. Errors. Tolerances.
Statistical and probability concepts applied to errors. Lifetime
measurements. Accuracy and sensitivity of indicating
instruments. Significant figures in a measurement.

2   Electrical measurements   26
Moving coil, moving iron and dynamometer instruments. Voltmeters, ammeters and wattmeters. Wattmeter corrections. The
cathode ray oscilloscope — operation and use. Measurements of
resistance, the Wheatstone Bridge. Measurement of capacitance
and inductance, the a.c. bridge. The d.c. potentiometer.
Magnetic flux measurement. Digital meters.

3   Measurement of mechanical quantities   60
Mass, length and angle. Tensile tests. Hardness and surface
finish. Torque measurement.

4   Measurements of other quantities   75
Temperature — alternative methods of measurement. Calorific
value of fuels. Light measurements — luminous intensity, the
photo-voltaic cell. Sound measurements — phons and decibels.
Sound in buildings — sound level instruments. Fluid measurements — pressure and flow. Measurement of time.

5   Electronics I (analogue circuits)   97
The thermionic diode. Semiconductors and the pn junction.
The transistor and its use in an amplifier. 'h'-parameters.
Characteristics and the use of the load line. Frequency
response.

6   Electronics II (logic circuits)   118
The four basic logic statements AND, OR, Exclusive OR and
NOT gates. Truth tables. Linked logic gates. Boolean algebra
rules. Fluidic logic gates — other fluidic devices.

7   Transducers   136
Strain gauges. Velocity measurements, linear and angular.
Acceleration measurement. Pressure transducers. Piezo electric
effect. Sensors.

8   Signals and data transmission   147
    Concept of information, transmission of information. Require-
    ments of a transmission system. Analysis of signals. Amplitude
    modulation. Multiplexing. Signal delays and signalling speeds.
    Signal distortion, noise, reflections. The binary signal. Error
    detecting codes. Remote position indicators.

    Index   167

Chapter 1

General principles of measurement

1.1 Standards and fundamental units

Before we can make any measurement it is necessary to have an acceptable standard unit on which to base the measurement. This standard unit has to be recognised both nationally and internationally and has to be measured to a high degree of precision. If, for example, we wish to measure the distance between two points in London, we use as our standard of length the **Metre** (or its subdivision the millimetre) and we know that our measurement will be accepted both in Manchester and Melbourne. At one time in Great Britain we used the yard as our standard of length, this being based on some fairly arbitrary units evolved from medieval times. The original standard was set up in Greenwich in 1855 and was the distance between marks on a bronze bar (Fig. 1.1).

Apart from the fact that we have now adopted a metric system the increasing precision often required nowadays in the measurement of length

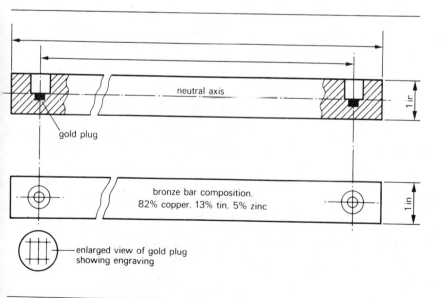

Fig. 1.1   The original British standard of length.

makes the bronze bar an insufficiently accurate primary standard. The bar is subject to variation of length with changes of temperature; but what is perhaps more unacceptable is its shrinkage – it has been contracting at a fairly uniform rate of a millionth of an inch per year. Today we require a standard which does not change, which is as precise as possible, and which can be fairly readily set up. We have turned to the physical phenomena of light radiation as our primary standard of length. Each electron contained within a single atom has been found to possess a certain very well defined amount of energy. When the energy of an electron is made to change from one precise level to a lower one, the change of energy is given up in the form of an electromagnetic radiation of an extremely well defined wavelength. Sometimes the radiation falls in the visible spectrum. This is an inherent property of the atom itself and is therefore of a fundamental nature. This provides the basis of our standard of length.

The standard unit of time is also based on a radiation measurement.

In all there are seven standard units, although outside chemistry only six are normally employed. They form the basis of the Système Internationale employed in Europe and many countries outside Europe. The term Système Internationale has been abbreviated to SI and we now refer to the SI units.

1.2 Basic SI units employed in engineering

(a) Length

The primary unit of length is the **Metre** and is based on the radiation wavelength of the gas krypton 86 (an isotope of krypton) in a vacuum. Under certain conditions the gas emits orange light of a very precise wavelength.

The metre is defined as 1,650,763.73 wavelengths of this particular radiation (Fig. 1.2).

The fact that nine significant figures are quoted suggests the degree of precision on which we now base the metre. It is closer to 1,650,763.73 wavelengths than 1,650,763.72 wavelengths, for example.

Originally the distance from the North Pole to the equator along the meridian through Paris was intended to be 10,000,000 metres. The present-day standard is fractionally longer.

(b) Mass

The primary unit of mass is the **Kilogramme** and is the mass of a platinum-iridium alloy cylinder held in the International Bureau of Weights and Measures at Sèvres near Paris. It has not been possible to obtain a more fundamental measurement of mass.

(c) Time

The primary unit of time is the **Second**. Originally it was defined in terms of the mean solar day (1/86400 part of a day) but present demands of

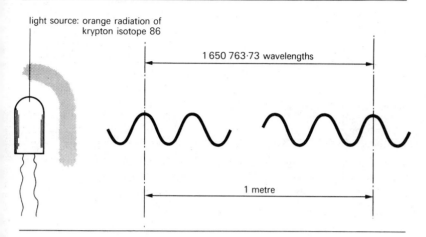

Fig. 1.2 The light standard of length.

accuracy have forced us into a more precise definition. Again we turn to the radiation phenomena for our standard.

The second is now defined as the time for 9,192,631,770 cycles of the radiation from vaporised caesium under resonant conditions. There are ten figures in this definition and once again there is an implied very high degree of precision.

(d) Other units

There are three other primary units employed in engineering — units of electric current, temperature and luminous intensity. The unit of electric current is the **Ampere** and is defined as follows:

The ampere is that unvarying electric current which, when flowing through two, infinitely long, straight, parallel conductors of negligible cross section and placed 1 metre apart in a vacuum produces a force of 2×10^{-7} newtons per metre length of conductor (Fig. 1.3).

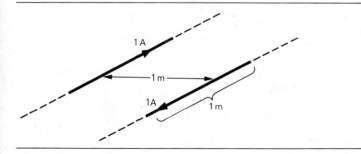

Fig. 1.3 Definition of the ampere.

The metre has already been defined and since the newton can be measured in terms of length, mass and time it follows that the definition of the ampere is dependent upon other SI units.

The SI unit of thermodynamic temperature is the **Degree Kelvin (K)**. This is directly related to the Celsius (or Centigrade) scale. The scale is such that the temperature difference between absolute zero (0 K) and the triple point of water (0.01°C), where water can be simultaneously in the solid (ice), liquid and gaseous (vapour) state, is 273.16 K.

On this basis

Zero degrees Kelvin 0 K = −273.15°C

The unit of luminous intensity is the **Candela** and is defined in terms of the light output from a unit surface area of molten platinum.

The final SI unit is the **Mol**, seldom employed by engineers but used extensively by chemists.

Table 1.1   SI units

Quantity	Unit	Symbol
length	metre	m
mass	kilogramme	kg
time	second	s
electric current	ampere	A
temperature	kelvin	K
luminous intensity	candela	cd

Two other units must also be mentioned. Both are non-dimensional and both are used in angular measurement. The first is the **Radian**. An angle in radians is defined as follows:

$$\text{Angle in radians} = \frac{\text{Arc length at radius } r \text{ subtended by the angle}}{\text{radius } r}$$

Hence the angle in radians of a circle is

$$\frac{2\pi r}{r} = 2\pi \text{ radians}$$

The other is the **Steradian** or solid angle. This is the three-dimensional equivalent of the radian. The angle at the top of a cone is an example of a solid angle. It is defined by

$$\text{Solid angle in steradians} = \frac{\text{Area at radius } r \text{ subtended by the angle}}{(\text{radius } r)^2}$$

Hence the angle in steradians of a sphere is

$$\frac{4\pi r^2}{r^2} = 4\pi \text{ steradians}$$

1.3 Sub-standards or secondary standards

The custodian of standards in this country is the National Physical Laboratory (N.P.L.), Teddington (now supplemented by the British Calibration Service), but it would be highly inconvenient if we had to check every measurement against the primary standards held there. It is much more usual for various organisations which require standards to produce their own standards or equipment which may then be checked against the primary standards periodically. These sub-standard or secondary standards as they are called are usually in a rather more convenient form than the N.P.L. standards, e.g. a sub-standard ammeter which has been calibrated against the standard ampere is much easier to use in subsequent measurements. Likewise a metrology department in industry would hold a number of 'slip gauges' which are essentially length standards and these would be used to check the accuracy of working gauges or micrometers which are in daily use. A slip gauge is made to very precise limits and depending upon its size might be accurate to 1 micron (1 μ) (a micron is 10^{-6} m). (See Fig. 1.4.)

It is necessary to monitor the accuracy of working standards employed in production in a factory since the quality of the final product is dependent upon them. Whenever there is a significant fall in the accuracy or quality of a component coming off the production line one of the checks which must be made is the working standard. The frequency at which

Fig. 1.4 Metric slip gauges.

standards are checked is a matter for debate and depends to a large extent on the particular standard. It is more likely that the calibration of a deflection type ammeter may change over a period of 6 months than, say, the dimensions of a slip gauge. It is vital that if the quality of a product is to be maintained then not only should a factory check its standards regularly but full records should be kept of such calibration so that long-term changes in working standards are apparent. The 'traceability' of standards calibration is vital.

1.4 Calibration errors

The calibration of an instrument or gauge is normally carried out in a standards laboratory where there is some control over the environment, e.g. temperature and humidity which might affect the calibration. The instrument to be calibrated is compared against the standard and a graph is drawn of 'error' against 'indication'.

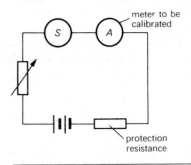

Fig. 1.5   Calibration of an ammeter.

For example, the calibration of an ammeter involves a simple circuit (Fig. 1.5) consisting of a sub-standard ammeter S in series with the meter to be calibrated A. The current through both meters (which is the same) is adjusted by means of the variable resistor. The current is set to perhaps one-fifth of the full-scale deflection on A and the sub-standard indication is noted on S. A typical graph is shown in Fig. 1.6. Occasionally readings are taken for both ascending and descending values of current. There will normally be allowable limits of error laid down for the instrument and if the calibration curve lies within these limits over the entire range (5 A in this case) then the instrument is passed as an 'industrial grade' meter or 'precision grade' meter. It should be noted that although the calibration points lie within the acceptable limits (see Fig. 1.6), the meter is only accurate at zero, 2.4 A, 3.3 A (and then only as accurate as the sub-standard). When the meter indicates 1.0 A there is an error of +0.05 A, i.e. the correct indication should be 0.95 A. Such errors are referred to as

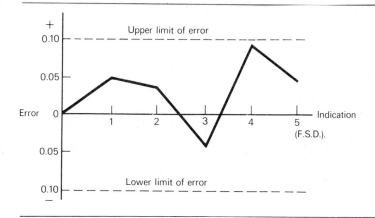

Fig. 1.6   Error graph.

calibration errors. In the example given the allowable error is ±0.10 A at all points of the scale. As a percentage error on the indication the allowable error at 1.0 A is ±10 per cent of the indication. At full scale deflection (5 A) the allowable error is still ±0.1 A, which is now ±2 per cent of the indication. It follows that as far as indicating pointer instruments are concerned it is advantageous to arrange matters so that readings are taken as near full scale deflection as possible. Whatever instruments are used, whether electrical or mechanical, they are always subject to a calibration error. The allowable error is usually stipulated in a British Standard specification (BS) and these errors should be known before a measuring instrument is used. The allowable error for certain types of industrial grade pointer ammeters is given by BS 89 and is ±2 per cent of full scale deflection. If when a meter or instrument is checked its calibration is found to be outside the standard specification then it should be either recalibrated or returned to its manufacturer. Often the reason for the error being outside allowable limits is due to some malfunction of the instrument so that when this happens a careful check should be made of its operation.

1.5 Systematic errors

Systematic errors may be attributable to the method of measurement employed. When measurements are made the system which is being measured is disturbed by the instruments which are used. The amount of disturbance should normally be made as small as possible or at least allowed for. Take for example the measurement of a resistance by the voltmeter—ammeter method (Fig. 1.7). If the voltmeter is connected to point 'a' the voltmeter reading ÷ ammeter reading gives the resistance of the resistor + the resistance of the ammeter in series. If the voltmeter is connected to point 'b' the current flowing through the ammeter is that

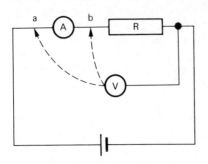

Ammeter 0–5 A, Industrial grade (2%).
Resistance 1.7 Ω.
Voltmeter 0–100 V, Industrial grade (2%).
Resistance 100 kΩ.

Fig. 1.7   Measurement of resistance.

through the resistor + the current flowing through the voltmeter. The ratio of the voltmeter and ammeter readings will not give the true value of the resistance for either method of connection. In either case it is possible to calculate the systematic error. The calibration errors of the two instruments will result in further sources of error.

Example 1.1

A resistor of about 20 Ω is to be measured as shown in Fig. 1.7 by the ammeter–voltmeter method. Details of the two meters are given. A reads 3.0 A, V reads 60 V. What are the probable limits in the measurement of the resistance?

The voltmeter has an error of ±2 per cent of full scale deflection (F.S.D.). The voltage recorded would therefore be between 60 + 2 per cent of 100 and 60 − 2 per cent of 100, i.e. between 62 V and 58 V. Similarly the ammeter would have calibration errors lying between ±2 per cent of 5 A, i.e. ±0.1 A. The ammeter indicates a value between 3.1 A and 2.9 A.

Taking the worst cases, the observed value of resistance lies between

$$\frac{62}{2.9} = 21.3 \ \Omega \quad \text{and} \quad \frac{58}{3.1} = 18.7 \ \Omega$$

The voltmeter is, however reading the voltage across the resistor and the ammeter and this introduces a systematic error. The resistance given by the ratio of voltage and current includes the ammeter resistance (1.7 Ω). This value must be subtracted from the above values so that the probable value of R lies between 19.6 Ω and 17.0 Ω. Note that in this measurement the systematic error could have been reduced by connecting the voltmeter to point b. The high value of the voltmeter resistance would increase the current flow through the ammeter only fractionally. No higher degree of precision would have been achieved because the systematic error has already been taken into account.

Systematic errors may be due to environmental conditions. A steel cylindrical gauge being used to measure a diameter in a low temperature cabinet has contracted and its calibration is slightly different to that at room temperature. If the temperature of the cabinet and the temperature at which the gauge was calibrated are known then with the further knowledge of the coefficient of expansion of steel the systematic error can be calculated and allowed for.

1.6 Observational and random errors

Under this heading we can place all the other forms of measurement error. There is firstly the personal error of observation. Given a deflectional type instrument where it is necessary to interpolate between markings on a scale, different observers obtain slightly different readings for the same deflection. Sometimes there may be a gross observational error — a figure of 45.6 may be wrongly observed as 55.6 or we may be reading the wrong range on a multirange instrument. The increased use of digital meters reduces the possibility of observational errors. Even however when all the calibration systematic and observational errors have been taken into account a residual error remains which cannot be accounted for since there is a certain degree of randomness in all measurements. The random error should be small but becomes evident when measurements are repeated. Almost invariably slight changes in observed values occur. It is seldom possible to repeat the observed indications precisely. To take one example, a digital instrument will always be liable to a ± 1 digit random error in any measurement.

1.7 The economics of measurement

Accurate measurement costs time and money and it would be quite wrong to use sophisticated and costly equipment making a precise measurement where an approximation would do. The speedometer in a car for instance need not be very accurate and no one would be unduly worried by a 5 per cent calibration error. The air speed indicator on a large passenger-carrying aircraft must be more precise since there may be danger of approaching stalling speed without the pilot knowing it and catastrophic results might occur with a 5 per cent uncertainty in air speed. There has to be an awareness of the degree of precision which the circumstances demand and it is wasteful to produce measurements to an unnecessary degree of accuracy.

1.8 Accuracy and sensitivity

The accuracy of an instrument is basically the limitations on its indication within which we cannot place any reliability. The sensitivity or discrimination on the other hand controls the smallest change in its indication which can be discerned. Sensitivity can be expressed as

$$\frac{\text{Change of indication (output)}}{\text{Change of measured quantity (input)}}$$

e.g. in a pressure gauge the sensitivity might be twenty divisions per N/m^2. The sensitivity is normally required to be high so that small changes in the quantity being measured produce relatively large changes in indication. High sensitivity does not necessarily imply high accuracy but high accuracy usually requires high sensitivity.

For instance, an instrument may have a calibrated scale on which we can discern a change of 0.1 per cent of its full-scale deflection but the instrument may be only accurate to ±10 per cent of its full-scale deflection due to a fault.

In general the more sensitive an instrument is, the more likelihood there is of it also being more accurate, but there is no guarantee of this.

A Wheatstone bridge uses a 'null balance' to obtain a measurement (see p. 45). The sensitivity of the bridge depends upon the smallest change in resistance value of one of the four arms of the bridge which causes a discernible current to flow through the galvanometer. If the bridge balances with the variable arm set to 1759 Ω and we know that the accuracy of the resistance is ±0.2 per cent then we might at first glance say that the value of the unknown resistance was between 1759 Ω ± 0.2 per cent, i.e. between 1762 Ω and 1756 Ω. However, if we require a change of 15 Ω in the setting of the resistance before we can detect a change of current the sensitivity of the bridge is such that we cannot say with any certainty where the balance point is between the limits 1759 ± 15 Ω i.e. 1774 Ω and 1744 Ω.

Very often there is a lower limit placed on accuracy in the appreciation that there is a limit to the accuracy which an instrument or standard can be made.

A micrometer for example may be stated to have an accuracy of ±0.02 per cent of the reading ±0.01 mm error. If we measure a cylinder of diameter 100 mm with this instrument the accuracy would be

100 ± 0.02 per cent ± 0.01 mm

i.e. the upper limit would be 100.03 mm
the lower limit would be 99.97 mm.

If the same micrometer were used to measure the diameter of a 1 mm cylinder the 0.02 per cent error (0.0002 mm) would be quite negligible in comparison with the ±0.01 mm error (the lower limit).

For this measurement the limits of accuracy would be

1 ± 0.01 mm
i.e. 1.01 mm and 0.99 mm.

As a percentage of the nominal value (1 mm) this is ±1 per cent and not

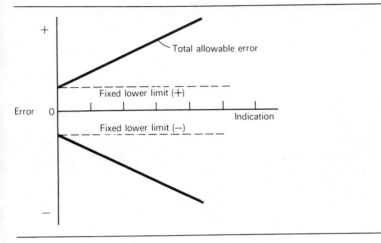

Fig. 1.8   Lower limit of accuracy shown graphically.

±0.02 per cent as implied by the stated accuracy of the micrometer.

The effect of this lower limit of accuracy can be shown graphically in Fig. 1.8.

1.9   Total error and tolerances

It should be obvious by now that when making a measurement we ought to be in a position of recognizing the sources of error and computing the total error or the confidence limits we can place on the results. We can express the result as the most probable value together with the limits of accuracy,

e.g. 15.67 mm ± 0.25 mm or 25.8 kg ± 2 per cent.

Although we normally expect the total error to be the same amount either side of the stated value we have some situations in which an error lies more probably on one side of a given value than the other.

A plug gauge (Fig. 1.9) is used to check the diameter of a circular hole. If it has a diameter of 50 mm then any hole it fits into merely has a

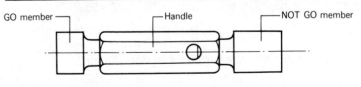

Double-ended plug gauge

Fig. 1.9   Double-ended plug gauge.

diameter greater than 50 mm. Indeed, since the plug gauge itself has a tolerance and cannot be made precisely 50 mm we must assume that the holes into which it fits are larger than the minimum value of the plug gauge's diameter.

Thus if the 50 mm is itself subject to a tolerance of ±0.001 mm, for example, there is a chance that the gauge is only 49.999 mm diameter and we can only say with complete confidence that the holes into which it fits have a diameter greater than 49.999 mm. The lower limit is therefore zero and the upper limit is unspecified.

Similarly the holes which would not take the gauge would have an upper limit of 50 mm nominally, but again we could only state with complete confidence that if the plug gauge did not fit into the hole then its diameter would be less than 50.001 mm.

When fitting parts together it is necessary to put tolerances on the limits of accuracy required and the tolerance allowed is often given as a figure with only negative or positive errors (not both).

If for example a number of cylinders are to be a tight fit in a number of tubes then the inside diameters of the tubes cannot be smaller than the outside diameters of the cylinders. If the fit must be between certain limits we must use the following type of statements:

(*a*) The cylinder must be 15 mm outer diameter − 0 mm − 0.05 mm.
(*b*) The tube must be 15 mm inner diameter − 0 mm + 0.10 mm.

The tolerance on the fit between the cylinder and tube for the worst possible case is 0.15 mm. Many fits would be much better than this. Because of the calibration error of the gauges used to measure the diameters of tube and cylinder there may be a possibility of a non-fit − i.e. the cylinder diameter being fractionally larger than the tube diameter. This is however somewhat unlikely. This possibility could be removed by choosing gauges such that even allowing for their calibration errors the lower limit on the gauge used for the inner diameter of the tube was no smaller than the upper limit on the gauge used for the diameter of the cylinder.

Where holes or diameters are to lie between certain limits we employ 'GO' and 'not-GO' plug gauges. The diameter of the 'GO' end of the gauge (see Fig. 1.9) represents the lower acceptable diameter and the 'not-GO' end of the gauge represents the higher acceptable limit. In order that the holes in a component are acceptable (i.e. between specified limits) the 'GO' gauge must be capable of being pushed into the hole, while the 'not-GO' gauge should not fit.

1.10 Measurement of large numbers of components

The situation often presented in an industrial organisation is the measurement of a large number of components which are supposedly identical but show a scatter about some value.

Let us consider the manufacture of about 2000 cylinders which are nominally 20 mm diameter. There will be some variation with many cylinders about this value. Few will be precisely 20 mm diameter. The results might be something like Table 1.2.

Table 1.2

Average diameter	No. of cylinders
19.82	14
19.87	63
19.92	154
19.97	291
20.02	746
20.07	327
20.12	191
20.17	78
20.22	40
20.27	5
	$\bar{x} = 20.03$

A histogram shows more clearly the scatter around the required diameter (Fig. 1.10).

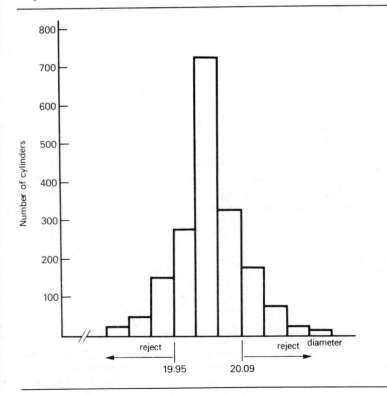

Fig. 1.10   Histogram.

If the acceptable limits are 19.95 mm and 20.09 mm then the number outside these limits is:

14 + 63 + 154 + 191 + 78 + 40 + 5 = 545

The number within acceptable limits is 291 + 746 + 327 = 1364. If the total number of measurements is increased and the interval between readings decreased, the stepped histogram curve becomes a smooth curve (Fig. 1.11). Furthermore if the possibility of measurements occurring below or above the mean value is the same, that is the errors about a mean value are purely random, then the ideas of probability can be introduced.

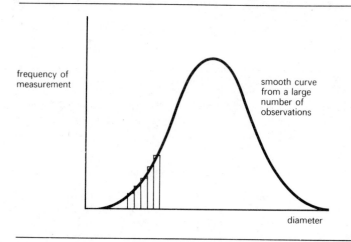

Fig. 1.11   Distribution graph.

The spread of observations about an average value can be expressed in a number of ways but the most usual scientific method is to use the concepts of 'mean' and 'standard deviation'.

The mean value of a number of readings (n) is the sum of all the readings divided by the number of readings taken.

$$\text{Mean} = \frac{\text{Sum of readings}}{\text{number of readings taken}} = \frac{\Sigma x}{n}$$

(The symbol Σ means the 'sum of'.)

Taking the example already given, the mean value of each diameter range is half-way between minimum and maximum values of that range, i.e. the cylinders in the range 19.95 and 19.99 have a mean value of 19.97 (perhaps more strictly 19.975 since the next step starts at 20.00). We have therefore 291 cylinders with an average diameter of 19.97 mm.

The mean diameter of *all* the cylinders is found by taking the total of all the diameters measured. We therefore have to take the mean of each diameter range and multiply by the number of readings in each range and repeat this for each range.

$$\text{The mean of all the cylinders} = \frac{\Sigma(\text{av. dia. for each range} \times \text{no. of cylinders in each range})}{\text{total no. of cylinders}}$$

$$\text{Mean} = \frac{(14 \times 19.82) + (63 \times 19.87) + (154 \times 19.92) \dots + (5 \times 20.27)}{1909}$$

$$= 20.03 - \text{call this } \bar{x}$$

The deviation of a particular measurement x from the mean is $x - \bar{x}$ and obviously this can be positive or negative depending upon whether x is greater or less than $\bar{x}$.

The standard deviation (S.D.) of n measurements is defined mathematically by

$$\text{S.D.} = \sqrt{\frac{\Sigma(x - \bar{x})^2}{n}}$$

symbol σ

Example 1.2

What is the S.D. (σ) of the following twenty readings?

17.1	18.2	17.3	16.7	19.9	19.7	18.5
18.2	18.9	19.1	17.0	20.3	19.6	18.8
18.6	18.0	18.1	17.7	18.9	20.1	

$n = 20$

$$\text{The mean value} = \frac{17.1 + 18.2 + 17.3 \dots 18.9 + 20.1}{20}$$

$$= 18.54 \; \bar{x}$$

The evaluation of the S.D. can be best evaluated in tabular form (see Table 1.3). From this $\sigma = 1.03$.

The question might well be asked at this juncture. What is the point of all this?

Briefly we can say that the standard deviation gives a mathematical measure of the amount of scatter of a number of readings about a mean value — small S.D. — small scatter. We must have a sufficient number of observations to make this meaningful otherwise we may come to false conclusions. The mean of a small sample and the corresponding S.D. may be different from the values obtained with a large sample. When the measurements are given in the form of a histogram we evaluate the S.D. in the following manner:

$$\text{S.D.} = \sqrt{\frac{\Sigma(\text{Area of a narrow strip or step}) . (x - \bar{x})^2}{\text{Total area under the curve}}}$$

Table 1.3

Reading	x	$x - \bar{x}$	$(x - \bar{x})^2$
1	17.1	−1.44	2.074
2	18.2	−0.34	0.116
3	17.3	−1.24	1.538
4	16.7	−1.84	3.386
5	19.9	1.36	1.850
6	19.7	1.16	1.346
7	18.5	−0.04	0.002
8	18.2	−0.34	0.116
9	18.9	0.36	0.130
10	19.1	0.56	0.314
11	17.0	−1.54	2.372
12	20.3	1.76	3.098
13	19.6	1.06	1.124
14	18.8	0.26	0.068
15	18.6	0.06	0.004
16	18.0	−0.54	0.292
17	18.1	−0.44	0.194
18	17.7	−0.84	0.706
19	18.9	0.36	0.130
20	20.1	1.56	2.434

$$\bar{x} = 18.54$$

$$\Sigma(x - \bar{x})^2 = 21.294$$

$$\frac{\Sigma(x - \bar{x})^2}{n} = 1.0647 \quad \text{S.D.} = \sqrt{\frac{\Sigma(x - \bar{x})^2}{n}} = 1.031$$

Again, taking the example of the cylinders the mean value $\bar{x}$ and the standard deviation can be found from a suitable table (Table 1.4).

Thus the mean value of diameter is 20.03, the standard deviation is 0.0743 and the range of observations is 0.5 mm (19.8 − 20.3). If the chances of a diameter being greater or less than the average are even the histogram will by symmetrical about $\bar{x}$ and will approach a curve called a Gaussian distribution. We can then predict the chances of a particular cylinder diameter being within tolerance limits. There is the assumption that the observations are made on a typical sample and we cannot be sure of this. The reliability of our prediction increases as the number of observations increase.

On the basis of this assumption we may then say that the Gaussian curve has equal areas inside and outside limits when the deviation either side of the mean is 0.6745 × standard deviation. This is the limit at which the chance of a cylinder diameter being greater or less than this amount is the same.

At a deviation = ±1 S.D. the fraction of the area within limits is 0.6826. The chances here are 0.68 to 0.32 or about 2 : 1 that a particular diameter is within these limits. At two standard deviations the fractional area is 0.9546, i.e. chances of about 0.95 to 0.05 or 19 : 1. At 3 S.D.s the fraction

Table 1.4

a	b	c = (a − x̄)	c²	d = b × 0.05	c² × d
Av. dia.	No. of cylinders	x − x̄	(x − x̄)²	Area in strip	
19.82	14	−0.21	0.0441	0.70	0.03087
19.87	63	−0.16	0.0256	3.15	0.08064
19.92	154	−0.11	0.0121	7.70	0.09317
19.97	291	−0.06	0.0036	14.55	0.05238
20.02	746	−0.01	0.0001	37.30	0.00373
20.07	327	+0.04	0.0016	16.35	0.02616
20.12	191	+0.09	0.0081	9.55	0.07736
20.17	78	+0.14	0.0196	3.90	0.07644
20.22	40	+0.19	0.0361	2.00	0.07220
20.27	5	+0.24	0.0576	0.25	0.01440
				Total area = 95.45	Σ = 0.52735

Mean $\bar{x}$ = 20.03

$$\sigma = \sqrt{\frac{0.52735}{95.45}} = 0.0743$$

hence approximately two-thirds of the cylinders have diameters between 20.104 and 19.956. ($\bar{x} \pm \sigma$).

has increased to 0.9972, i.e. 99.7 per cent of all the measurements fall inside the limits. (See Fig. 1.12.)

In the example taken the tolerance limits on the mean value of 20.03 mm would have to be ±0.149 mm in order to predict a 19:1 chance that the cylinder diameters would be within tolerance, i.e. only 1 in 20 would be rejected.

It must be emphasized that all this is based upon the assumption that the results follow a random pattern which will produce a Gaussian distribution. If, due to tool wear, the cylinders measured are not a typical sample then the predictions will not hold.

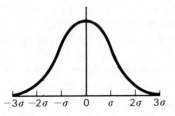

Fig. 1.12 Gaussian distribution.

1.11 Lifetime measurements

This work becomes of increasing importance where large numbers of identical components are produced and where a reliability or lifetime is to be found.

Take for example the prediction of the lifetime of electric filament lamps. If we wish to guarantee the life then we have to make measurements on a sample to ensure that the life exceeds the guaranteed minimum. Unfortunately we have to test a lamp to destruction before we can say with any certainty what its lifetime is and in this case it is necessary to test only random samples. The optimum size of the sample depends upon many factors which are partly mathematical and partly economic, i.e. the cost of making the measurement (and possibly destroying the product as in the case of lifetime measurements), the reliability required in the final result and the time available in which to make the measurements. Suppose a sample of n items is taken from a large group of products which have a Gaussian distribution about some average value. If the average value $\bar{x}_1$ of the sample is taken it will approximate to the average value of the original group of products, but may be slightly more or slightly less. If the average value of another sample of n items is taken it will also have an average value $\bar{x}_2$ which again will approximate to the average value of the large group. If the process is repeated a number of times the mean value of $\bar{x}_1$, $\bar{x}_2$, etc., should equal the average value of the original group. Moreover it can be shown that the values $\bar{x}_1$, $\bar{x}_2$, etc., follow themselves a Gaussian distribution and have a standard deviation of $\sigma/\sqrt{n}$ where σ is the standard deviation of the whole group and n is the number in each sample taken. This ratio is called the standard error of the mean. For example, let us assume that we have a large number of components, say 10,000, which have an average value of 500 and a standard deviation of 35. If we took a number of different sample batches each 25 in number then the standard deviation of the average values of each sample would be $35/\sqrt{25} = 7$, while the mean value of the average values of the samples taken would be 500 (very nearly). It is obvious that the larger the number in each of our samples the smaller the standard deviation, i.e. the closer does the sample resemble the total group.

The problem facing the statistician is of the sort: how large must the sample be if the probability of its average value lying not more than a given amount from the mean of the total is not to exceed a certain value?

In the example given, if there must be a 19 : 1 chance of the mean of the sample lying between 490 and 510, how big must the sample be?

To solve this type of problem it is necessary to have recourse to probability tables (Table 1.5).

If the chances are 19 : 1 this means 5 per cent of the cases must lie outside the limits of 490 and 510, or 2½ per cent are greater than 510 and 2½ per cent smaller than 490. Taking the upper limit we get

$$\bar{x} + \frac{\sigma}{\sqrt{n}} \times p = \bar{x} + x_1$$

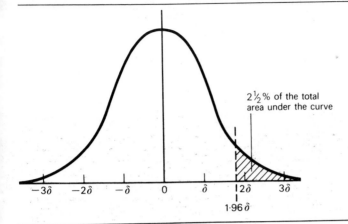

Fig. 1.13 Normal probability curve.

Table 1.5 Area under the normal probability curve

	0.00	0.01	0.02	0.03	0.04	0.05	0.06	0.07	0.08	0.09
0.0σ	0.5000	4960	4920	4880	4840	4801	4761	4721	4681	4641
0.1σ	0.4602	4562	4522	4483	4443	4404	4364	4325	4286	4247
0.2	0.4207	4168	4129	4090	4052	4013	3974	3936	3897	3859
0.3	0.3821	3783	3745	3707	3669	3632	3594	3557	3520	3483
0.4	0.3446	3409	3372	3336	3300	3264	3228	3192	3156	3121
0.5	0.3085	3050	3015	2981	2946	2912	2877	2843	2810	2776
0.6	0.2743	2709	2676	2643	2611	2578	2546	2514	2483	2451
0.7	0.2420	2389	2358	2327	2296	2266	2236	2206	2177	2148
0.8	0.2119	2090	2061	2033	2005	1977	1949	1922	1894	1867
0.9	0.1841	1814	1788	1762	1736	1711	1685	1660	1635	1611
1.0σ	0.1587	1563	1539	1515	1492	1469	1446	1423	1401	1379
1.1	0.1357	1335	1314	1292	1271	1251	1230	1210	1190	1170
1.2	0.1151	1131	1112	1093	1075	1056	1038	1020	1003	0985
1.3	0.0968	0951	0934	0918	0901	0885	0869	0853	0838	0823
1.4	0.0808	0793	0778	0764	0749	0735	0721	0708	0694	0681
1.5	0.0668	0655	0643	0630	0618	0606	0594	0582	0571	0559
1.6	0.0548	0537	0526	0516	0505	0495	0485	0475	0465	0455
1.7	0.0446	0436	0427	0418	0409	0401	0392	0384	0375	0367
1.8	0.0359	0351	0344	0336	0329	0322	0314	0307	0301	0294
1.9	0.0287	0281	0274	0268	0262	0256	0250	0244	0239	0233
2.0σ	0.0228	0222	0217	0212	0207	0202	0197	0192	0188	0183
2.1	0.0179	0174	0170	0166	0162	0158	0154	0150	0146	0143
2.2	0.0139	0136	0132	0129	0125	0122	0119	0116	0113	0110
2.3	0.0107	0104	0102	0099	0096	0094	0091	0089	0087	0084
2.4	0.0082	0080	0078	0076	0073	0071	0070	0068	0066	0064
2.5	0.0062	0060	0059	0057	0055	0054	0052	0051	0049	0048
2.6	0.0047	0045	0044	0043	0041	0040	0039	0038	0037	0036
2.7	0.0035	0034	0033	0032	0031	0030	0029	0028	0027	0026
2.8	0.0026	0025	0024	0023	0023	0022	0021	0021	0020	0019
2.9	0.0019	0018	0018	0017	0016	0016	0015	0015	0014	0014
3.0σ	0.00135									

where $\bar{x} + x_1$ = the upper limit and p = the reading off the table corresponding to an area of 0.0250 (i.e. 2.5 per cent) under the probability curve (Fig. 1.13).

If the chance factor is 0.025 then from the table $p = 1.96$.

$$\therefore 500 + \frac{\sigma}{\sqrt{n}} \times 1.96 = 510$$

or $\sqrt{n} = \dfrac{\sigma \times 1.96}{10}$

Since $\sigma = 35$ $\therefore \sqrt{n} = 6.86$

$$n = 47 \text{ to the nearest integer.}$$

In round figures we must choose about fifty random components from the total of 10,000 in order that we have about a 20 : 1 chance in our favour that this sample has a mean lying between 490 and 510.

1.12 Poisson distribution

When dealing with sampling methods in reliability or lifetime measurements use is often made of the Poisson distribution. The probability of an event occurring is given by the respective terms in the expression

$$e^{-m}\left(1 + m + \frac{m^2}{\underline{2}} + \frac{m^3}{\underline{3}} \dots \right) \text{ where } m = \text{the occurrence rate.}$$

To illustrate by way of an example, suppose in a large factory that on average a fluorescent tube fails every 1.67 hours. The failure rate is $\dfrac{1}{1.67} = 0.6$ tubes/hour. The probability of no tubes failing in an hour is given by the first term in the expression

i.e. e^{-m}.   For $m = 0.6$ $e^{-m} = 0.549$

the probability is 549 : 1000 or roughly 11 chances in 20.

Thus in 20 *separate* hours taken at random on 11 occasions there will be no lamp failure. In the other 9 there will be one or more lamp failures. One must of course remember that this is a probability only based on statistical theory.

Taking the second term

$e^{-m} \times m$.   For $m = 0.6$ $me^{-m} = 0.329$

or a chance of about 33 : 100 that one lamp will fail.

Likewise taking the third term

$e^{-m} \times \dfrac{m^2}{\underline{2}}$. For $m = 0.6$ this gives 0.099

The chances are 99 in 1000 or about 1 in 10 that two lamps will fail in a given hour.

One can add the probability figures. Thus 0.329 + 0.099 = 0.428. So that there is a chance of 428 to 1000 of one or two lamps failing in an hour.

Taking a second example based on the cylinders once more. Let us assume that in 2000 cylinders 16 are outside tolerance. If the cylinders are stacked in groups of 200 what are the chances that there will be 3 or more outside tolerance in each stack?

The number outside tolerance in each stack on average is $\dfrac{200}{2000} \times 16 = 1.6$. This is the occurrence rate.

Taking the Poisson series:

$e^{-1.6} = 0.202$ This is the probability of zero cylinders being outside tolerance.

$1.6 \times e^{-1.6} = 0.323$ This is the probability of 1 cylinder being outside tolerance.

$\dfrac{1.6^2}{\underline{2}} \times e^{-1.6} = 0.258$ This is the probability of 2 cylinders being outside tolerance.

Adding $\overline{\underline{0.783}}$ represents the probability of 2 or less being outside tolerance in a stack.

Hence

$1 - 0.783 = 0.217$ is the probability of 3 or more being outside tolerance in a stack of 200, i.e. very roughly the probability is 1 stack in 5 will contain 3 or more cylinders outside tolerance limits.

1.13 Significant figures

If a measurement is made there must be a certain confidence in the result. The unit of length has been quoted to nine significant figures and the implication is that we can have confidence in all the nine figures quoted, but rarely is it possible to put more than two or three significant figures down in a laboratory report with any real meaning. We must always remember that in a series of measurements it is the least accurate one which usually controls the confidence that we can place on the final result and therefore affects the number of meaningful significant figures. Although one does not easily fall into the trap of quoting a pointer instrument reading to five significant figures there is a temptation to do so when using digital instruments and one must always be on one's guard. For example, the current through a circuit may be found by measuring the voltage across a resistance using a digital voltmeter. It is pointless measuring this voltage to five significant figures if the resistance is known only to an accuracy of 5 per cent.

On the other hand, if a small difference between two quantities is to be measured then the more significant figures in the two measurements the better, providing that the significant figures do not stray beyond the accuracy of the individual indications. A pressure drop may be measured by noting the levels of two mercury manometers across an orifice. If the two levels are 98 mm and 95 mm then the difference is 3 mm. If the two levels can be measured to a greater accuracy, e.g. 98.3 mm and 95.1 mm, then the difference is 3.2 mm. If however little reliance can be made on the figures after the decimal point in the two initial readings no reliance at all can be placed on the figures after the decimal point in the difference measurement and it is misleading to quote the difference as 3.2 mm. (This is illustrated in the next section.)

One other situation which may produce nonsensical results is the use of electronic calculators.

The stress in a wire may be measured directly using the relationship

$$\text{Stress} = \frac{\text{Load}}{\text{Area of cross section}}$$

If the load is subject to ±0.5 per cent error and the area to ±2 per cent error the total error is ±2.5 per cent.

It is quite ridiculous to use an electronic calculator to calculate the ratio and quote all the figures indicated in the readout (often seven significant figures). If the stress was 96.9 units then the error would be such that the result would be between 99.4 and 94.4, i.e. only the first two figures are meaningful and not too much reliance can be placed on the second.

1.14 Some philosophical points

A number of different methods may be available when making a measurement and the choice of the method employed will naturally depend upon the measuring apparatus to hand, and the degree of precision required. Generally speaking a method involving a continuous measurement will be superior to one in which discrete measurements are made. For example, the temperature gradient in a length of piping carrying a hot liquid can be found by monitoring continuously the temperature difference over a given length of piping rather than making 'spot checks'.

This will reduce environmental errors as the temperature-sensing devices used will have had ample time to have reached the temperature of the liquid. Moreover the temperature of the liquid will no longer be affected by the introduction of the devices. Any small perturbations in the temperature gradient will also have been averaged out.

The actual method of making the measurement can normally be chosen from a variety, for instance in the case of temperature measurement the choice lies between the use of resistance thermometers, thermocouples or mercury thermometers. If a continuous recording is required an electrical sensing device would often be preferred.

If the temperature gradient in the example given is small then special precautions would have to be taken since the measurement would have to rely on the difference between two measurements.

A numerical example will serve to illustrate.

Let the higher temperature t_1 be $50°C$ and the lower temperature t_2 be $47°C$. Then the difference will be $3°C$. However, if the accuracy of each measurement can be guaranteed only to within $0.5°C$ (i.e. approximately 1 per cent) the difference temperature can show wide variations about the nominal figure of $3°C$. Taking the worst cases, if $t_1 = 50.5°C$ and $t_2 = 46.5°C$ there is $4°C$ difference and if $t_1 = 49.5°$ and $t_2 = 47.5°$ the difference is only $2°$.

This means that there is a difference of $\pm1°$ about $3°C$ — a considerable percentage variation.

So that the confidence limits on the *difference* measurement are $3°C \pm 33$ per cent, whereas each measurement has accuracy limits of about ±1 per cent. It may not be possible to measure other than by a 'difference' measurement in some cases. In these instances the measuring devices must be very precise to ensure a reasonable accuracy on the difference result.

Summary

Measurements must be based on standards. In the SI units these standards are: length — metre; mass — kilogramme; time — second; electric-current — ampere; temperature — degree kelvin and luminous intensity — candela.

Errors occur when making measurements and these can be categorized as Calibration errors, Systematic errors and Random errors. Accuracy of a particular measurement is usually dependent upon the sensitivity or discrimination of the apparatus used but great sensitivity does not necessarily produce great accuracy.

It is necessary to state total error and tolerance in many industrial situations. When a large number of nominally identical components are being manufactured certain laws of statistics may be employed to monitor the quality of a product. Sampling may be used in such cases. The concepts of mean, standard deviation and standard error are used. The Poisson series may be employed in certain instances. Precautions are especially necessary when a measurement depends upon the small difference between two quantities. Care must be exercised when quoting results to more than two or three significant figures.

Questions

Who is the custodian of standards in this country?

What sort of accuracy is expected in a surface finish?

What precautions are necessary when using slip gauges?

How frequently should standards in a technical college laboratory be checked? Consider two examples: (i) the metrology laboratory and (ii) an electrical laboratory.

Examples

1. A pressure gauge is known to be accurate to ±1.5 per cent of the full scale deflection (1000 N/m²). An observer notes a reading of 860 N/m² and his observational error is ±1 division in 200 on the scale. Between what limits does the pressure lie?
Ans. 880 and 840 N/m²

2. The stress in a circular cross section wire under tension is measured directly. The load is subject to an error of ±0.2 per cent. The diameter is measured with an accuracy of ±0.05 per cent. What is the total error and how many significant figures can be quoted with meaning in the stress?
Ans. ±0.3 per cent   Three significant figures

3. Using Kater's pendulum 'g' is given by

$$g = \frac{4\pi^2(h_1 + h_2)}{T^2}$$

The length $h_1 + h_2$ was measured as 1042.4 ± 0.5 mm and the time of oscillation T as 2.049 ± 0.0005 s.
Calculate the maximum possible percentage error in 'g'.
Ans. ±0.055 per cent
(*Kent Colleges OND (Tech.)*)

4. In a test on a component in manufacture a failure occurs on average once every 2000 hours. Using the Poisson Distribution calculate

(i) the probability of more than one failure in a period of 1000 h
(ii) the probability of no failures in 4000 h
Ans. 0.0903 and 0.1353
(*Kent Colleges OND (Tech.)*)

5. The breakdown voltages for twenty specimens of insulation film are as follows:

Voltage kV	4.2	4.3	4.4	4.5	4.6	4.7
Frequency	1	3	7	5	2	2

Find the mean and standard deviation voltages.
Ans. 4.45 kV   0.128 kV
(*Kent Colleges OND (Tech.)*)

6. The current in a circuit is given by $I = V/R$. If $V = 200$ V with an error of +0.1 V and R is 100 Ω with an error of −0.2 Ω calculate the error in I in mA.
Ans. +5 mA
(*Kent Colleges OND (Tech.)*)

7. The power dissipated by a resistor is measured by using a voltmeter and ammeter. The voltmeter is placed across the resistor and ammeter in series. The voltmeter reads 135 V and its F.S.D. is 200 V. The ammeter reads 0.90 A and its F.S.D. is 1.0 A. Find the apparent power and the limits within which the power really lies. The instruments are both accurate to 1 per cent of F.S.D. and the resistance of the voltmeter is 10,000 Ω and the ammeter is 1.50 Ω. Observational errors may be neglected.
Ans. 121.5 W,   123.5 W and 116.2 W.

8. (*a*) For a particular null method, the unknown value 'X' is related to the known adjustable value 'S' by $X = k \cdot S$ in which 'k' is dimensionless. Show that the limita-

tions of detector sensitivity and known-value discrimination impose opposite requirements on the value of 'k'.

(*b*) A Wheatstone bridge, having a ±0.005 per cent six-decade 0—1 ohm to 0—100 kilohm standard resistor and ratio resistors of 10, 100 and 1000 ohms, is used to measure a resistance of approximately 1000 ohms. Assuming the bridge ratio to be exact, evaluate the measurement uncertainty for the available ratios and hence select the ratio which sensibly minimises the uncertainty and requires a not-too-sensitive detector.

Ans.

k	Per cent uncertainty	
0.01	0.00505	
0.1	0.0055	
1.0	0.010	'Best' $k = 0.1$
10.0	0.055	
100.0	0.505	

(*Kent Colleges OND (Tech.)*)

Chapter 2

Electrical measurements

2.1 Introduction

Most electrical measurements involve the measurement of one or more of the following quantities:

Current, Voltage, Power, Phase angle (between two sinusoids of the same frequency), Frequency, Resistance, Capacitance and Inductance.

We may also wish to observe the waveform of voltages and currents.

In the past pointer instruments have usually been employed in many electrical measurements, but today, with the availability of cheap and adaptable integrated electronic circuits, an instrument with a digital display is often only marginally more expensive than the pointer instrument. Digital instruments suffer one major defect — they need electrical power for operation and therefore, before they can be used, a mains or battery supply must be available. They are also complicated (and can therefore go wrong). On the credit side they are robust, accurate, easily read and often sensitive. In spite of their sensitivity they can withstand electrical overloads. They are also becoming increasingly reliable. While there will always be a substantial need for the pointer instrument, industry is employing more and more digital display instruments.

2.2 The basic pointer instruments

These are mainly of three types: moving coil, moving iron and dynamo-meter.* Each meter requires three torques for its operation.

1. A deflecting torque. This is produced by the interaction of two magnetic fields. The quantity being measured is responsible for producing at least one of the magnetic fields.
2. A control torque. This can be produced by a spiral of beryllium copper (Fig. 2.1) or by a taut ribbon (Fig. 2.2).
3. A damping torque. This reduces the oscillations of the pointer about its average deflection.

In each instrument the deflecting torque acts against the action of the control torque and when these are equal no further deflection occurs. The damping torque disappears once a steady deflection has been reached.

*Dynamometer instruments are also referred to as electrodynamic meters.

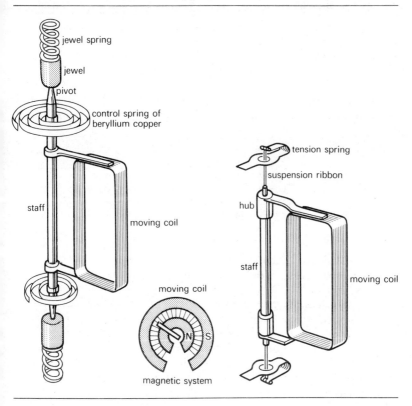

Fig. 2.1　The moving-coil meter.　　　　**Fig. 2.2**　Taut ribbon control.

Each instrument has a full-scale deflection (F.S.D.) and a sensitivity related to the current (or voltage) required for F.S.D. The full-scale deflection obviously depends upon both the deflecting torque and the control torque at full scale.

2.3　The moving-coil meter

The moving-coil meter relies on the relationship:

$F = BlI$ where F is a force in newtons

l is a length in metres

I is the current in amperes

and B is a magnetic flux density in Teslas (Wb/m²)

This force occurs when a current-carrying conductor is placed in a magnetic field perpendicular to the flux lines. If a coil is placed in a magnetic field with circular magnetic poles, as in Fig. 2.1, which create a radial field then the total force on the coil $F = BIl \times N$, where N is the number of turns. The torque $T = BIlN \times r$, where r is the effective radius of the coil and I is the current in the coil.

The deflecting torque produced by this current causes the coil to twist which tightens the control spring. The control torque is usually directly proportional to the angle turned through, i.e. $\propto k\,\theta$. When the control torque and deflecting torque are equal and a steady deflection is obtained then

$$k\theta = BIlNr$$

or $\theta = \dfrac{BIlNr}{k}$

Providing B, l, N, r and k remain constant:

$\theta \propto I$,

i.e. the deflection is directly proportional to the current.

This means that a linear scale results and the spacing of the scale divisions is equal.

If the flux density is non-uniform the scale becomes non-linear.

The damping torque is provided by winding the coil on an aluminium former. As the coil moves in the magnetic field an e.m.f. is induced in it and an eddy current flows in the former itself. This produces a retarding torque dependent upon the angular velocity of the coil. This damping effect disappears when the final deflection has been reached.

Moving-coil meters can be manufactured to give a wide variation of full-scale deflection. The more sensitive of them have a full-scale deflection of only a few microamperes. The very sensitive types are referred to as galvanometers. These have a very small control torque and in consequence

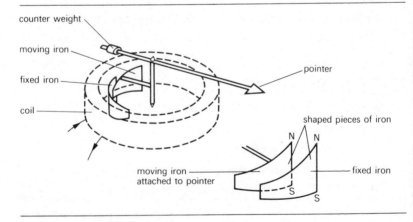

Fig. 2.3   Moving-iron meter (repulsion type).

are rather delicate and susceptible to any vibration. They tend to be limited to use in a laboratory situation.

2.4 The moving-iron meter

The principle of this meter relies on the force produced between two pieces of iron when magnetised by a magnetic field. When current flows through the coil (Fig. 2.3) the two shaped pieces of iron become magnetised in the same direction and repulsion occurs between the pairs of north poles and the pairs of south poles. One of the iron pieces is fixed and the other is attached to the pointer. Consequently the pointer experiences a deflecting torque.

This torque is directly dependent upon the square of the number of turns (N) on the coil and the current (I) flowing in the coil, i.e.

deflecting torque $\propto (NI)^2$

As with the moving-coil meter, control and damping torques are needed. The control torque is provided in precisely the same way as previously, but this time the deflecting torque is produced by current flowing in the stationary coil. Damping is usually provided by an air dashpot (Fig. 2.4).

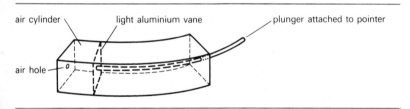

Fig. 2.4 Air dashpot.

The air in the chamber is compressed by the movement of the pointer from its zero position. The degree of damping is controlled by the size of the hole in the dashpot.

The scale shape is affected by the shape of the fixed- and moving-iron pieces. Careful design can make the scale linear over some 80 per cent of its length. It must be remembered that the deflection is dependent upon I^2.

2.5 Dynamometer instruments

The third type of indicating meter is the dynamometer instrument, which in some respects is similar to a moving-coil meter. It consists of a moving coil suspended between suitable bearings and connection is made to it via control springs. The arrangement is similar to that used in the moving-coil meter but the coil is wound on a non-metallic former.

The permanent magnet of the moving-coil meter is replaced by two fixed coils which when passing current produce a magnetic field (see Fig. 2.5). The fixed coils are normally air cored, but there are some dynamometer instruments available with iron cores.

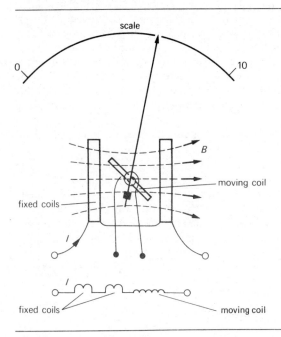

Fig. 2.5   Dynamometer; ammeter or voltmeter connection.

The fixed coils may be connected in series with each other and in series with the moving coil, so that the same current passes through each. The flux density B produced by the fixed coils depends upon the current I flowing:

$B \propto I$

The torque produced can be found from the same relationship as that employed in the moving-coil meter, i.e.

$$T = BIINr$$

Thus

Torque $\propto IIIN$

$\propto I^2$, since l and N are constant.

The control torque is provided by the helical springs and damping is provided by an 'air dashpot'.

Dynamometer ammeters are often employed as precision instruments but they are not as sensitive as permanent-magnet moving-coil meters.

Care must be exercised when stray magnetic fields are present in the vicinity of the meter as these can produce a torque on the moving coil and hence affect the indication. The meter can be employed as a voltmeter by the addition of a series resistor but by far the greatest use of the instrument lies in its ability to measure power.

2.6 The dynamometer wattmeter

The construction is similar to that of the dynamometer ammeter except that the connections to the fixed coils are brought out independently as are the connections to the moving coil. The moving coil is connected in series with a resistor.

The torque produced is proportional to the product of the flux density B provided by the fixed coils and the current i in the moving coil.

B depends upon the current I (Fig. 2.6)
i depends upon the voltage V across the moving-coil circuit.

Hence

torque $\propto Bi$
$\quad\quad\propto IV$ which is the power

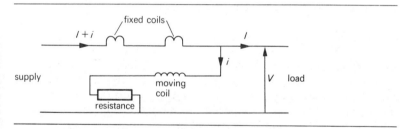

Fig. 2.6 Dynamometer; wattmeter connections.

This result (torque $\propto$ power in the circuit) applies both to d.c. and a.c. circuits. In the a.c. case the indication is proportional to $IV \times \cos \varphi$, where φ is the phase angle between V and I. The fixed-coil connections are arranged so that the two coils may be connected either in series or parallel. With parallel connections the current flowing through each coil is half of that in a series connection. Hence the flux density B is reduced by a factor of two. This halves the torque and hence the indication of the meter. The reading obtained for power will then have to be multiplied by 2 to obtain the correct value.

The way the coils are connected is shown in Fig. 2.7. The series resistance to the moving coil is often tapped at various points to accommodate

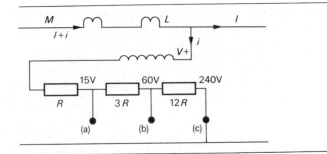

Fig. 2.7   Multirange wattmeter.

various voltages, and once again this affects the calibration of the meter. For example a voltage V may produce a current i in the moving coil when connected to the lowest tapping point (a on the diagram). But a voltage $4 \times V$ is needed to produce the same current i, when connected to the second tapping point (b on the diagram).

The meter indication in the second case must be multiplied by 4 to obtain the true power. For the meter shown in Fig. 2.7, the multiplying factors shown in Table 2.1 must be used.

The F.S.D. of the instrument in each case is 15 W.

The decision on which connections to use depends upon the voltage and current in the circuit. Obviously if I is less than 1 A, the series connection of the current coils would be employed. For currents between 1 A and 2 A either mode of connection could be employed as the current coils are normally designed to withstand overloads, but obviously the nearer the current is to 2 A the better it is to employ the 2 A connection (current coils in parallel). A similar argument applies to the voltage circuit.

For alternating currents in excess of 2 A it is necessary to employ a current transformer which will reduce the current to an acceptable value. The current ratio will then involve a further multiplying factor. A potential transformer can be employed in the voltage circuit in a similar manner. Current and potential transformers can only be used in a.c. circuits.

Table 2.1

Current coils	Voltage tapping point	Max. power	Multiply reading by
1 A (in series)	15 V	15 W	1
1 A (in series)	60 V	60 W	4
1 A (in series)	240 V	240 W	16
2 A (in parallel)	15 V	30 W	2
2 A (in parallel)	60 V	120 W	8
2 A (in parallel)	240 V	480 W	32

Example 2.1

A wattmeter is used to measure the power in a 250 V a.c. circuit and Table 2.1 is relevant.

The current coils are in parallel (max. current 2 A) and a 5/1 current transformer is used. The voltage across the circuit is connected to the 240 V tapping point in the voltage circuit. The wattmeter indicates 10.5 W (F.S.D. 15 W). What is the power dissipated?

The 240 V tapping point and the 2 A connection implies a multiplying factor of 32. There is an additional factor of 5 for the current transformer. Hence true power

= 10.5 x 32 x 5
= 1680 W

2.7 Wattmeter corrections

When a wattmeter is used to measure power, systematic errors are introduced because of the method of connecting the instrument into the circuit.

The power dissipated in the case of d.c. circuits is $I \times V$, where I = current through the load resistor and V = voltage across it.

It can be seen from Fig. 2.7 that the voltage coil of the wattmeter is across the load, but the current flowing through the current coils consists of the load current I plus the current i through the voltage coil.

The indication on the wattmeter depends on the voltage across the voltage coil multiplied by the current through the current coils. The wattmeter reading is therefore

$V(I + i)$

$= VI + Vi$

= power in the load ⏌ Vi

The reading is therefore high by the amount Vi which is the power dissipated in the voltage coil. When the load current is small this error can be appreciable. The true power in the load can be found by subtracting this power from the wattmeter reading. Since

$V = iR_v$, where R_v = voltage coil resistance

$Vi = i^2 R_v$

or

$Vi = V^2/R_v$

These are alternative ways of determining the power dissipated in the voltage coil.

The current coil may of course be connected on the other side of the voltage coil (see Fig. 2.8a). Although only the load current I is now

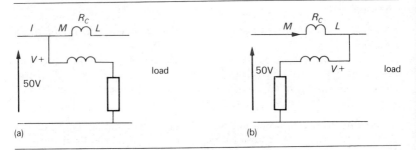

Fig. 2.8   Two methods of connecting a wattmeter to a load.

flowing through the current coil, the p.d. across the voltage coil is V (the voltage across the load) plus IR_c (the p.d. across the current coil-resistance R_c).

Indication on the wattmeter $= (V + IR_c)I$
$$= VI + I^2R_c$$
$$= \text{power in the load}$$
$$+ I^2R_c$$

The reading is therefore high by the amount I^2R_c which is the power dissipated in the current coil. (The error can be appreciable when V is small.)

The true power in the load can be found by subtracting this power from the wattmeter reading. These corrections also apply when the supply is an alternating one.

Example 2.2

Find the power in a load given that the wattmeter reading is 55 W when the direct supply voltage is 50 V, the wattmeter being connected with its voltage coil across the supply.

The current coil resistance is 6 Ω and the voltage coil 1200 Ω. What would be the wattmeter indication for the alternative method of wattmeter connection?

The 50 V is across both current coil and load resistance. Here the 55 W is the total power in the load and the current coil (see Fig. 2.8a).

$I = 55/50 = 1.1$ A
Power in the current coil $= I^2R_c = 1.1^2 \times 6$
$$= 1.21 \times 6$$
$$= 7.26 \text{ W}$$
True load power $= 55 - 7.26 = 47.7$ W

With the alternative method of connection (see Fig. 2.8b) the voltage across the load can be found as follows:

The load resistance = power/I^2 = 47.7/1.21 = 39.3 Ω

The voltage across the load is the same as the p.d. across the voltage coil resistance and this is in series with the p.d. across the current coil resistance. The combined resistance of 39.3 Ω and 1200 Ω in parallel is 38.0 Ω. Hence the p.d. across the voltage coil and the load resistor in parallel (see Fig. 2.8b)

$$= \frac{50 \times 38.0}{38.0 + 6} = 43.1 \text{ V}$$

Power in the load   = $43.1^2/39.3$ = 47.5 W

Power in the voltage coil = $43.1^2/1200$ = 1.55 W

Therefore the indication on the wattmeter = 49.05 W.

It can be shown that the wattmeter reads the true average power in a.c. circuits. The systematic errors often become more important where alternating currents are concerned because the power dissipated is dependent not only upon the product of current and voltage as in the d.c. case but also on the cosine of the phase angle between them. When the angle approaches 90° the power in the circuit may be small even with comparatively large values of current.

2.8 Ammeters and voltmeters

There is no basic difference between ammeters and voltmeters. Both are essentially current devices, but the former are characterised by a low resistance, the latter by a high resistance.

An ammeter is usually arranged in parallel with a resistance, the 'shunt', which diverts a large part of the current, and leaves only a proportion of the current to flow through the meter itself. The shunt must therefore be of a low resistance compared with the meter resistance. Since it may carry a large current it must be so constructed that it does not become overheated. It must also have a resistance which does not change with temperature.

With some meters the shunt is an external component which must be fitted to the instrument, whilst with others a switch enables various shunts to be connected inside the meter. The ohmic value of the shunt affects the full-scale indication of the meter.

Example 2.3

A meter of resistance 48 Ω has a full-scale deflection of 1 mA. What values of shunt resistance are required in order that the F.S.D. should be (a) 10 mA, (b) 5 A?

(a) Since the meter gives full-scale deflection with 1 mA, then under the shunted conditions 9 mA must flow through the shunt.

The p.d. across the meter is then 1×48 mV, so the p.d. across the shunt must likewise be 48 mV.

Hence the shunt resistance =

$$\frac{V_{shunt}}{I_{shunt}} = \frac{48 \times 10^{-3}}{9 \times 10^{-3}} = 5.334 \ \Omega$$

Similarly (*b*): At F.S.D. current through the shunt

= 4.999 A

$$R_{shunt} = \frac{V_{shunt}}{I_{shunt}} = \frac{48 \times 10^{-3}}{4.999} = 0.00961 \ \Omega$$

The value of the shunt resistance in this case is small, and care must be taken to reduce any additional resistance presented by the connecting leads between the meter and shunt. Usually when an external shunt is used four terminals are provided, one pair specifically for connecting the resistance to the meter (see Fig. 2.9).

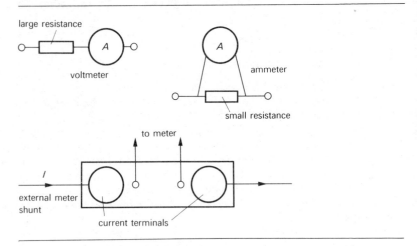

Fig. 2.9   Ammeters and voltmeters.

Example 2.4

A meter of 5 mA F.S.D. has total resistance of 25 Ω. How can it be converted to a voltmeter reading up to 100 V?

The meter can be converted by adding a suitable resistance in series. The value of this resistance must be such that when 100 V is connected across the meter and series resistor together, 5 mA flows.

Hence total resistance $(R_{meter} + R_{series}) = \dfrac{100}{5 \times 10^{-3}}\ \Omega$

$$= 20{,}000\ \Omega$$

Hence series resistor $= 19{,}975\ \Omega$

The sensitivity of voltmeters is often expressed as the ratio of total resistance to voltage giving F.S.D. In the example just quoted, the sensitivity would be $20{,}000/100 = 200\ \Omega/\text{V}$.

2.9 Meters for use in alternating current circuits

A meter which gives a deflection proportional to (current)^2 will indicate root mean square values (R.M.S.) and can be used in both a.c. and d.c. circuits. It follows that the moving iron and dynamometer meters read R.M.S. values. The moving coil meter can be adapted to read on alternating current circuits by the use of a bridge rectifier (Fig. 2.10).

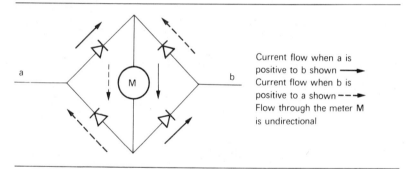

Fig. 2.10   Bridge rectifier.

It must be appreciated that it does not read r.m.s. values, and the d.c. calibration does not hold for a.c. circuits (except in the case of multimeters, where additional shunt resistors may be employed).

The indication on such a meter when connected to a sinusoidal supply is (peak value $\times\ 2/\pi$) where a bridge rectifier is used.

The ratio

$$\frac{\text{reading on alternating current}}{\text{reading on direct current}}$$

$$= \frac{\text{r.m.s. value}}{\text{average value of the full-wave rectified sine wave}}$$

$$= \frac{\text{peak value} \times 1/\sqrt{2}}{\text{peak value} \times 2/\pi}$$

= 1.11

This is known as the form factor.

Care must obviously be exercised in using meters and it must be appreciated that both waveform and frequency can affect the indication. It is essential to know the waveform and the frequency before deciding what reliance can be placed on the meter indication for any current or voltage. Table 2.2 summarises these points.

Table 2.2

Employed on	Moving-iron meter	Moving-coil meter	Rectifier instrument
d.c.	Correct reading	Correct reading	May give incorrect reading
50 Hz sine wave	Correct reading	Does not read	Correct if calibrated on sine wave
10 kHz sine wave	No use	Does not read	Correct if calibrated on sine wave
50 Hz non-sine wave	Reads r.m.s. if harmonic content small	May give an indication but not correct	Gives average value of waveform. May introduce large errors
10 kHz non-sine wave	No use	No use	Of doubtful use. Must be employed with caution

2.10 The cathode ray oscilloscope

The cathode ray oscilloscope (C.R.O.) is probably the most versatile and useful of all the electrical instruments in use today. It may be used in a variety of ways, to observe waveforms, to measure voltages and currents, to measure frequency, phase, and time, and to display information, e.g. numbers or pictures. The essential component in the C.R.O. is the cathode ray tube itself (see Fig. 2.11). It consists basically of an evacuated glass tube containing a heated cathode and a flat electrode, the anode, which has a small hole in its centre. A high potential difference between anode and cathode accelerates emitted electrons from the cathode. Some of these electrons pass through the hole and move with a high velocity along the neck of the tube, impinging on the flat screen at the end. The inside of the screen is coated with a material which fluoresces when the energy possessed by the high-velocity electrons is given up on impact. The electrode arrangement described can produce only a very ill-defined spot of light on the tube face. The size of the spot may be reduced, and its brightness increased, by the inclusion of further cylindrical electrodes in the neck of the tube (see Fig. 2.12). By suitable adjustment of the voltages

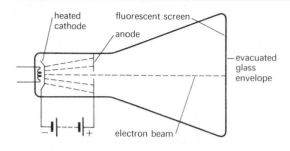

Fig. 2.11 Cathode ray tube.

applied to them, these electrodes produce an electric field which causes the electron beam to converge and come to a focus on the screen.

The beam can be further increased in intensity by the use of a negatively charged cylinder close to the cathode. This causes the electrons to converge towards the hole in the anode, so that a larger percentage of them pass through. The focusing of the electron beam is controlled largely by the potential applied to the second focusing cylinder. The intensity is controlled largely by the potential applied to the cylinder close to the cathode. It is possible, by making this electrode sufficiently negative, to cut off the electron beam completely.

Before we can make effective use of the tube it is necessary to add deflecting plates. These are pairs of plates situated either side of the electron beam in the vertical and horizontal planes.

A potential difference applied to the horizontal plates deflects the spot of light on the screen vertically, the electron beam moving towards the positive potential. Likewise, applying a potential difference to the vertical plates causes a horizontal electron beam movement. An alternating voltage will cause the electron beam to oscillate about a mean position and the spot on the screen will produce a straight line of light due to the combined effect of a slight 'after glow' in the fluorescent material and the eye's persistence of vision. The length of the line will be proportional to the peak-to-peak voltage applied to the plates.

The plates which cause vertical movement of the electron beam are referred to as the *Y* plates. The *X* plates produce horizontal deflection.

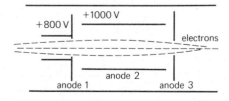

Fig. 2.12 Further electrodes.

The deflecting plates should be operated at roughly the same potential as the anode of the cathode ray tube, in order not to produce 'de-focusing'. In consequence it is usual to earth one of each pair of plates and also the anode. The heated cathode is then operated at a high negative potential with respect to earth in order to maintain the anode positive to the cathode, and thus accelerate the electrons from the cathode. The electrode supplies are shown in Fig. 2.13.

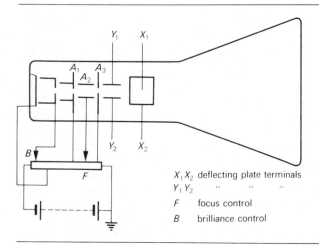

$X_1 X_2$ deflecting plate terminals
$Y_1 Y_2$ ''   ''   ''
F focus control
B brilliance control

Fig. 2.13 C.R.O. controls.

It is fairly obvious how the C.R.O. may be used to measure peak-to-peak values of voltages. The length of a line produced by a known sine wave of voltage can be used to calibrate the C.R.O. The deflection produced is directly proportional to the voltage even at frequencies up to 10 MHz or higher, since the effective mass of the electron beam is very small indeed. There is not the same inertia effect as would be present in moving-coil meters, for example.

If two sine waves which are of the same frequency are applied simultaneously to the X and Y plates then the pattern observable on the screen will vary between a straight line and an ellipse, depending upon the phase angle between the two sine waves. Figure 2.14 shows two cases, one of inphase conditions, and the other where a phase angle of 90° separates the two waves. Intermediate phase angles produce an ellipse whose major axis is inclined to the vertical. It is therefore possible to use the C.R.O. in this way to measure phase angles (but this method is not very accurate).

If sine waves of different frequencies are applied to the X and Y plates the pattern traced by the electron beam on the face of the tube can assume a variety of shapes. With a 2:1 ratio a figure of eight is produced. Patterns produced in this way are called Lissajous figures and two such figures are shown in Fig. 2.15.

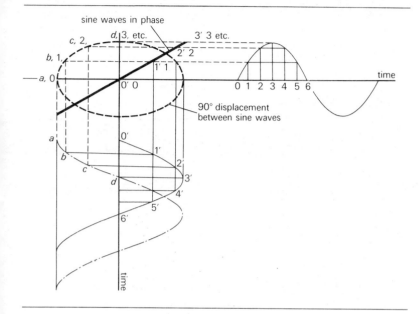

Fig. 2.14 Effect of phase angle on the display.

This is therefore a method of comparing or measuring frequencies, and can be used to calibrate a variable frequency source such as an oscillator.

2.11 The time base

If a waveform is to be observed on the screen of the C.R.O. then it is necessary to cause the spot of light to traverse the tube in the horizontal plane at a steady velocity while at the same time deflecting the spot vertically as the waveform being observed varies. This is achieved by applying the waveform under observation to the *Y* plates while applying a 'sawtoothed' waveform to the *X* plates. The latter is called a time base.

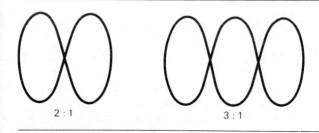

Fig. 2.15 Lissajous figures.

The sawtoothed waveform causes the spot to move from left to right in a linear manner while the X plate voltage is increasing. The spot returns to its original position when this voltage falls to zero. If the frequency of the sawtooth is precisely that of the waveform being observed, one complete wave is produced on the screen. If the timebase frequency is half that of the waveform then two complete waves are shown, and so on.

In practice the sawtooth voltage does not fall instantaneously to zero and a finite time elapses in which the spot returns to its original position (see Fig. 2.16). This time is known as the 'flyback' time, and it causes

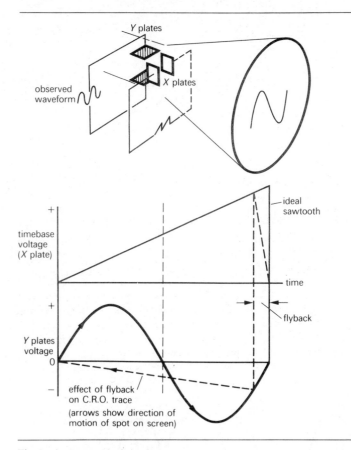

Fig. 2.16 Sawtoothed timebase.

some of the waveform to be lost. In most C.R.O.s the circuitry is arranged so as to 'black out' the trace during the flyback interval.

If the frequency of the timebase is not precisely that of the waveform (or a submultiple of it) then the trace on the screen appears to move either towards the left or the right. It is usually necessary to synchronize the time base. This is achieved by feeding a proportion of the observed wave-form to the circuit producing the sawtoothed wave. As a result, the build-up of voltage is slowed if the timebase speed is too fast, or conversely speeds it up if the timebase frequency is too low. It is necessary to ensure that the timebase frequency is nearly correct as the synchronization circuitry can cope only with small differences in timebase and observed waveform frequencies. This action is rather similar to the line hold control on a TV set.

The terminal on the C.R.O. to which this synchronizing voltage is fed is referred to as 'synch' and there is often a control on the oscilloscope which allows a variable proportion of the observed wave to be applied to the timebase.

The timebase just described is a 'free running' timebase. The tendency nowadays is to use a 'triggered' timebase.

The basic difference between them is that with a triggered timebase the observed waveform starts the timebase (that is, allows the timebase voltage to build up). At the end of the cycle the voltage falls to zero and the time-base is inoperative until the waveform starts it off again. This method allows a much greater control of the timebase and provides a steadier picture. It is also possible to expand the timebase to allow short-duration pulses to be observed.

Many oscilloscopes used today employ double beam tubes. These tubes have two independent focusing and deflecting assemblies enabling two traces to be shown simultaneously on the screen. Usually the X axis deflection is common to the two traces so that the two spots traverse the screen at the same velocity. This enables two different waveforms to be com-pared. The phase angle between two sine waves can be measured, for example, by superimposing one trace on top of the other.

C.R.O.s can cope with waveforms of high frequency (up to several hundred megahertz). The chief limitation is imposed by the frequency response of the amplifiers feeding the input signal to the deflecting plates.

2.12 Measurement of resistance

The measurement of resistance can be carried out in a number of ways and the choice of method is dependent upon the measuring equipment available, the accuracy required and the magnitude of the unknown resistance.

For low resistances (below about 10 ohms) the voltmeter–ammeter method described on p. 8 is often suitable. A known current is passed through the resistance and the potential drop it produces is measured. The error is in general the sum of the error of the instruments providing that the systematic error is allowed for. This method is frequently used in the measurement of the resistance of d.c. motor armatures where it is possible to pass appreciable current. One would not normally expect an accuracy better than about 4 per cent.

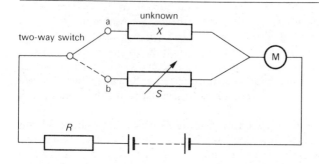

Fig. 2.17   Substitution method of measuring resistance.

For medium to high values of resistance — up to about 100,000 ohms — the substitution method may be employed. The circuit used is shown in Fig. 2.17. With the switch in position 'a' the current is noted. Resistance R is provided to protect the circuit from excessive currents. With the switch in position 'b' the current is adjusted by means of the calibrated resistance S until the meter M shows the same indication. The switch should be placed in position a to check that there has been no change of current. We can then say the X (the unknown resistance) = S (the calibrated resistance). Providing that we can detect a change in current on the meter we can set S to any degree of precision. The accuracy depends then on the accuracy of S, ignoring any resistance introduced by the switch. The method depends largely on having a meter with sufficient sensitivity in order to detect the smallest change in S (usually S is in the form of a decade resistance box).

This method is reasonably accurate and normally results with errors less than 1 per cent can be obtained.

The method most frequently employed in the measurement of resistance is the Wheatstone bridge (Fig. 2.18). The four resistances are referred to as the arms of the bridge, R_1 and R_2 being the ratio arms. R_3 is the variable usually calibrated in decades. R_x is the unknown forming the fourth arm. R_1, R_2 and R_3 are adjusted until the galvanometer G shows no current flowing when the key K is depressed. Under these conditions we say the bridge is balanced and no detectable current then flows through G.

It follows that the p.d. across R_1 must then equal the p.d. across R_x, i.e. $i_1R_1 = i_2R_x$. Likewise, the p.d. across R_2 must equal the p.d. across R_3, i.e. $i_1R_2 = i_2R_3$.

Collecting the two equations $\dfrac{i_1R_1}{i_2R_2} = \dfrac{i_2R_x}{R_3}$

$$\text{or } R_x = R_3\frac{R_1}{R_2}$$

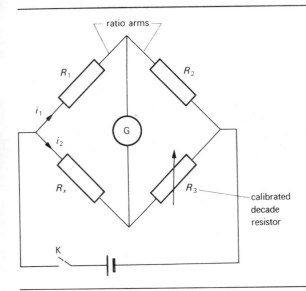

Fig. 2.18   The Wheatstone bridge.

If R_1 is 1000 Ω and R_2 is 100 Ω and R_3 is 137 Ω

$$R_x = \frac{1000}{100} \times 137 = 1370 \ \Omega$$

The arms R_1 and R_2, the ratio arms, are usually marked x (multiply) and ÷ (divide).

If the variable arm has four decades (i.e. units, tens, hundreds and thousands) it is possible to obtain a further significant figure in the measurement of R_2. If R_1 is set to 1000 and R_2 to 1000 then to obtain balance R_3 might have to be set to 1375 Ω to obtain balance, i.e.

$$R_x = \frac{1000}{1000} \times 1375 = 1375 \ \Omega$$

If the values of R_1 and R_2 have the values 1000, 100 and 10 Ω and R_3 can be set between values of 1 Ω and 9999 Ω then the largest resistance which can be measured is 999,900 Ω and the smallest is 0.01 Ω. The accuracy of the bridge depends upon the accuracy of the ratio arms and the calibrated arm and the sensitivity of the meter. If each arm is subject to an error of ±0.1 per cent the measurement will be liable to a possible cumulative error of ±0.3 per cent. At the extreme ends of the resistance range considerably greater errors would be encountered and we would be foolish to expect high accuracy outside the range 1 Ω and 10,000 Ω.

Precision Wheatstone bridges are made to high accuracy of probably better than ±0.02 per cent.

The accuracy of the bridge does depend upon having a galvanometer with sufficient sensitivity (see p. 10). It can be shown that maximum sensitivity occurs when the four arms are of about the same value of resistance. Many variations of the Wheatstone bridge exist but in each case the method of measurement depends upon obtaining a 'balance'. The bridge is a 'null method'. Self-balancing bridges exist where the 'out of balance' galvanometer current is used to operate a servo system which automatically changes the value of R_3 in a direction to restore balance. This method is used extensively in recording and controlling equipment and is shown diagrammatically in Fig. 2.19.

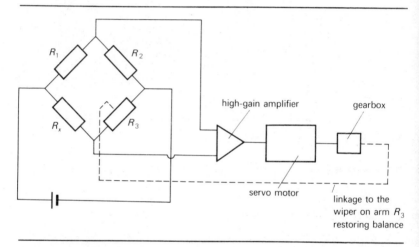

Fig. 2.19   Self-balancing bridge.

2.13 Measurement of capacitance

It is possible to measure large values of capacitance by the 'loss of charge' method. The capacitor is connected across a known high resistance, a high impedance voltmeter and a suitable d.c. supply as shown in Fig. 2.20. When the switch is opened the capacitor discharges through the resistor. The time taken for the voltage to fall to 37 per cent of its original value is the time constant CR.

In order that the time constant should be of a high value to ensure that the measurement of the time interval does not pose too many difficulties both C and R should be high. With R at 10 MΩ, a 1 μF capacitor will produce a time constant of 10 seconds which is about the minimum which can be measured with reasonable accuracy. This method therefore has its limitations. furthermore it is assumed that the leakage resistance of the capacitor itself is very much greater than the value R and the resistance of

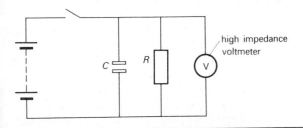

Fig. 2.20   Loss of charge method of measuring C.

the voltmeter is infinite. Although these can be allowed for by firstly measuring the time constant without the additional resistance this method is more of academic interest. One would be unlikely to obtain accurate readings from this method.

An alternative approach is to place the capacitor in an a.c. circuit and measure its reactance by observing the current through and the voltage across it:

$$\text{Reactance } X_C = \frac{\text{Voltage}}{\text{Current}} = \frac{1}{\omega C}$$

If the frequency is known then C can be calculated

$$C = \frac{\text{Current}}{2\pi f \times \text{Voltage}}$$

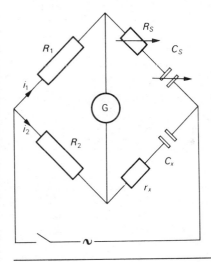

Fig. 2.21   Capacitor bridge.

Although rather better than the previous method where smaller capacitance is considered, three quantities — current, voltage and frequency — have to be measured. Nevertheless an accuracy of about 5 per cent may be obtained from this approach.

By far the best method is to place the capacitor in a Wheatstone bridge network as shown in Fig. 2.21. The two resistor arms R_1 and R_2 may be set to some arbitrary value (say 1000 Ω). The variable calibrated capacitor C_s in series with the variable calibrated resistor R_s are each adjusted until the galvanometer reads zero. When this occurs the potential drop across R_1 is equal to the potential drop across R_2.

$$i_1 R_1 = i_2 R_2 \tag{i}$$

It must be remembered that since we are dealing with a.c. quantities i_1 and i_2 have both magnitude and phase.

Similarly under balanced conditions the p.d.s across the two capacitor arms must be equal

$$i_1 Z_s = i_2 Z_x \tag{ii}$$

Z_x is the impedance offered by the capacitor we wish to measure. It must be appreciated that this capacitor may possess some resistance r_x which can be represented as a resistance in series with the capacitor.

Dividing (ii) by (i) we get

$$\frac{Z_s}{R_1} = \frac{Z_x}{R_2}$$

now $Z_s = R_s - \dfrac{j}{\omega C_s}$ (it is necessary to introduce the 'j' notation since impedance contains both resistive and reactive terms).

Similarly $Z_x = r_x - \dfrac{j}{\omega C_x}$

Hence $\dfrac{R_s - \dfrac{j}{\omega C_s}}{R_1} = \dfrac{r_x - \dfrac{j}{\omega C_x}}{R_2}$

This equation must be true for all conditions and therefore must be equally valid for the resistive (real) terms as for the reactive (j) terms. These terms can be equated separately.

Thus at balance

$$\frac{R_s}{R_1} = \frac{r_x}{R_2} \quad \text{and} \quad \frac{-j}{\omega C_s R_1} = \frac{-j}{\omega C_x R_2}$$

hence $r_x = \dfrac{R_2}{R_1} . R_s$ or $C_s R_1 = C_x R_2$

hence $C_x = \dfrac{R_1}{R_2} C_s$

Two conditions are necessary for balance in an a.c. Wheatstone bridge and this makes its operation that much more difficult than a d.c. bridge. The final balance equations in this case are independent of the frequency used.

The galvanometer in the a.c. bridge must be suitably designed to accommodate alternating currents. It can be a vibration galvanometer whose natural period of vibration coincides with the frequency of the supply voltage to the bridge. Nowadays such galvanometers have been largely superseded by high gain amplifiers feeding rectifier instruments.

The method described is reasonably precise and the accuracy depends upon the precision of R_1, R_2, R_s and C_s. A commercial form of this bridge gives an accuracy of about 1 per cent or better.

Many other forms of a.c. bridges exist. Figure 2.22 gives an example which can be used for measuring inductance.

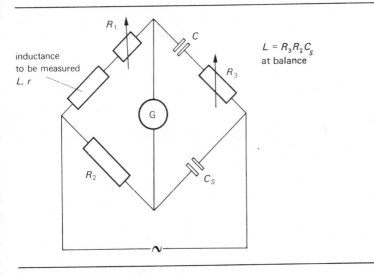

Fig. 2.22 Inductance bridge.

2.14 The potentiometer

The potentiometer is used mainly for very precise measurements of voltage. It has the advantage over the voltmeter of absorbing no current when making a measurement.

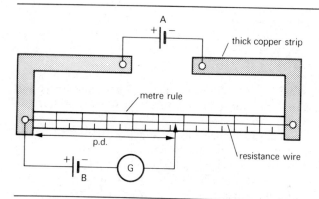

Fig. 2.23   Basic potentiometer.

Basically it consists of a thin uniform wire, usually a metre in length, to which is attached a tapping key connected to a galvanometer. The basic circuit is shown in Fig. 2.23. If we assume that battery A delivers exactly 2 V then the current which is drained from it due to the resistance of the slidewire produces a p.d. across a section of the slidewire proportional to the distance from one end. One metre represents 2 V, 50 cm represents 1 V, 1 cm represents 0.02 V (or 20 mV), and so on.

Battery B, with a terminal voltage less than 2 V, is now connected as shown, and the tapping key is adjusted until the galvanometer shows no deflection. The e.m.f. of B is exactly balanced by the p.d. across the section of the slidewire between the tapping key and the end connected to B.

Since 1 cm on the slidewire represents 20 mV, then, if balance occurs when the key is 55.6 cm from the end, the e.m.f. of cell B is 55.6 × 20 mV = 1112 mV or 1.112 V.

Since A may not give precisely 2 V across the slidewire, it is usual to 'standardise' the potentiometer before making a measurement. This is achieved using the circuit shown in Fig. 2.24 assuming that A supplies more than 2 V.

A small variable resistor R is connected in series with A. A standard cell, whose e.m.f. is known precisely, is then used in place of B. The tapping key is set to the position corresponding to the e.m.f. of the standard cell.

Suppose that the standard cell is a Weston cell. This is known to give an e.m.f. of 1.0186 V, or 1018.6 mV. As 20 mV is represented on the slidewire by 1 cm, 1018.6 mV is represented by 50.9 cm. The key is moved to this position and the circuit closed. If the galvanometer gives a deflection it indicates that the p.d. across the slidewire is not 2 V precisely, that is, the p.d. (from A) between the end connected to the Weston cell and the tapping point is not 1018 mV. The variable resistor R is now adjusted to change the current through the slidewire until the total p.d. across it is

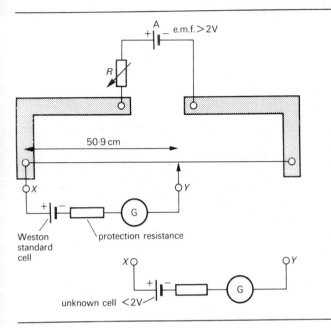

Fig. 2.24   Potentiometer with standardising resistance R.

2 V. This will be the case when the galvanometer shows no deflection. Providing there is no change in the current delivered from A there is no need to recheck the calibration for some time.

The standard cell is now replaced with the voltage to be measured, as shown in Fig. 2.24. The tapping key is adjusted for a new balance, and the p.d. calculated from the slidewire reading.

Commercial potentiometers are usually a little more complicated, but give a much more accurate result. Figure 2.25 shows a typical circuit. The galvanometer, standard cell and associated circuitry are built in. The network chain enables voltages to be set up very precisely, the slidewire now acting virtually as a 'fine setting' control. Protecting resistances are added to protect the galvanometer and standard cell from overloading.

Potentiometers are normally low-voltage devices, and if voltages in excess of 1 V are to be measured, additional potential dividing networks are needed so that a known proportion of the voltage to be measured is applied to the potentiometer. Like the Wheatstone bridge it is possible to obtain self-balancing potentiometers. The term potentiometer is also used to describe any three terminal variable resistor (Fig. 2.26). The voltage between one end and the movable tapping point depends upon the distance of the tapping point from one end. A variable proportion of the total voltage is available. The load current taken from the tapping point should be as small as possible.

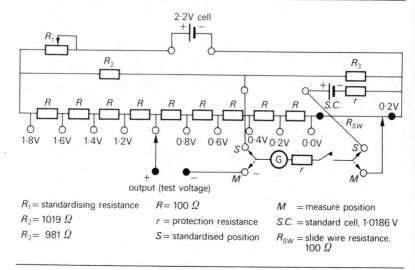

R_1 = standardising resistance R = 100 Ω M = measure position

R_2 = 1019 Ω r = protection resistance $S.C.$ = standard cell, 1·0186 V

R_3 = 981 Ω S = standardised position R_{SW} = slide wire resistance, 100 Ω

Fig. 2.25   Commercial potentiometer.

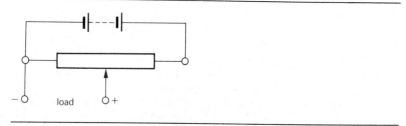

Fig. 2.26   Three terminal variable resistor acting as a potentiometer.

2.15 Measurement of magnetic flux

Several methods of measuring magnetic flux are available. Two of these measure essentially a *flux change*; the third method described can be used to measure a steady flux. Magnetic flux is measured in **Webers** or **volt-seconds** and this second unit gives a clue to one method of measurement.

When a magnetic flux linking a coil changes an e.m.f. is induced in the coil (Fig. 2.27). The magnitude of this voltage depends upon the number of turns N on the coil and the rate at which the flux changes $\dfrac{d\Phi}{dt}$. The relationship is

$$\text{e.m.f. } e \text{ (volts)} = N \frac{d\Phi}{dt}$$

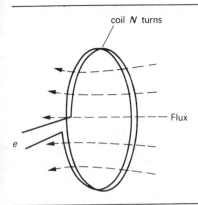

Fig. 2.27   E.m.f. induced in a coil.

The polarity of this e.m.f. is such that it tends to oppose the change of flux. (A number of text books show this opposition by the introduction of a minus sign in this expression.) The relationship can be rearranged to give

$$\Phi = \frac{1}{N} \int e \, \mathrm{d}t$$

Thus a flux change from Φ_1 to Φ_2 can be measured by integrating the e.m.f. e over the period of time that the change of flux occurs, i.e. measuring the area under the e.m.f./time graph. Whether the flux change occurs slowly or quickly is of no importance since the area under the e.m.f./time graph remains the same, i.e. we obtain a large e.m.f. for a short time, or a small e.m.f. for a longer time.

The integration of the e.m.f. can be carried out using an operational amplifier (Fig. 2.28). Here:

$$v_0 = -\frac{1}{CR} \int v_i \, \mathrm{d}t$$

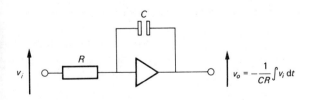

Fig. 2.28   Integrating amplifier.

Hence if the flux in a coil is to be measured the coil is connected to a current source via a switch. A second small coil (the search coil) is wound tightly over the first coil so that all the magnetic flux generated by the energised coil links the search coil. This is connected to an integrating operational amplifier and a suitable voltmeter. When the switch is opened the flux in the coil is reduced to zero (if no iron core is present). The flux change is then directly related to the voltage indicated.

If the time constant CR in the integrator is unity ($C = 1\ \mu\text{F}$ and $R = 1\ \text{M}\Omega$) then the flux change in the main coil is given by e/N Webers, where e is the voltage and N is the number of turns of the search coil. An adaptation of this circuit can be employed to produce a hysteresis loop (Fig. 2.29).

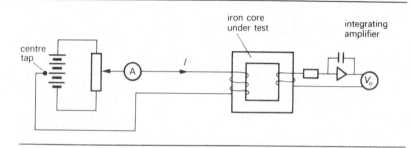

Fig. 2.29   Hysteresis loop apparatus.

If the iron core is initially in a demagnetised state and the wiper arm of the potentiometer is in the central position no current flows through the main winding. Displacing the wiper arm of the potentiometer from the central position causes a current to flow in the energising winding and produces a change of flux in the iron. This induces an e.m.f. in the search coil which is integrated and thus measures the flux change, the correspond-in magnetising current I being noted.

Plotting v_0 against I produces the magnetising portion of the graph. Pushing the wiper arm to the other end of the potentiometer slowly causes the current to reduce to zero and then reverse in direction.

Recording v_0 and I at suitable intervals and plotting gives the complete hysteresis loop. (See Fig. 2.30.)

A ballistic galvanometer which measures electric charge can be used to measure flux.

The change of indication on the meter is a measurement of the charge q (coulombs).

Now $q = i \times t$ (current $\times$ time in seconds)

or $dq = idt$

Integrating both sides of this equation $q = \int idt.$

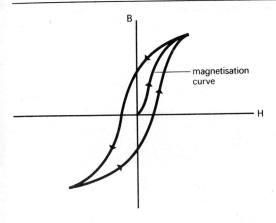

Fig. 2.30 Hysteresis loop.

If now the operational amplifier is replaced by a ballistic galvanometer and a resistance the galvanometer deflection will be proportional to the flux change. This is shown by the following.

Since $\Phi = \dfrac{1}{N} \displaystyle\int i\,dt$ and $e = ir$, where r = resistance of the search coil unit and i = current produced by the induced e.m.f. e.

Thus $\Phi = \dfrac{1}{Nr} \displaystyle\int i\,dt$

$\Phi = \dfrac{1}{Nr} \times$ change of deflection of the galvanometer.

Certain precautions are necessary in this type of measurement and the inductance of the search coil must be much smaller than r.

A meter to measure flux and operating on a similar principle is the flux-meter. It is designed such that its control torque is very small and its inertia relatively large. It can be shown that this overcomes the disadvantages of the ballistic galvanometer. The fluxmeter is calibrated in Weber turns and measures a *change* in flux. The change in deflection must be divided by the number of search coil turns to obtain the change of flux.

The third method of magnetic flux measurement employs the Hall effect. When a small block of semiconductor material is placed between the poles of a magnet and a current is passed between the faces of the semiconductor block an e.m.f. appears between the other two faces (see Fig. 2.31). The e.m.f. depends upon the current flowing, and the flux from the magnet, i.e.

$e = k\Phi I$

k is the Hall constant and it follows that

$$\Phi = \frac{e}{kI}$$

Hence if I is maintained constant the e.m.f. e is directly related to the flux Φ.

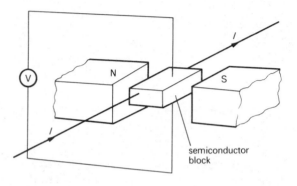

Fig. 2.31   Hall effect.

This is a direct measurement of flux rather than an indirect one since the relationship between Φ and e does not rely on flux change. Since the semiconductor block can be made very small it is possible to measure directly magnetic flux in small air gaps. If the area of the semiconductor block is known the average flux density over the surface area of the Hall probe can be found. It should be pointed out that the correct reading of flux will be obtained only if the magnetic flux is at right angles to the electrodes at which the e.m.f. is generated. The current fed to the probe is maintained constant by a suitable current generator.

2.16 Digital instruments

Digital instruments rely on the principle of converting an analogue quantity into a digital quantity. This process is called analogue to digital conversion (A to D) and is used in communication networks (see Chapter 8). The principle of operation is illustrated in Fig. 2.32. The input voltage V_{in} is first of all integrated and then applied to a comparator circuit and measured against a reference voltage. The reference voltage can be maintained at a known precise level. The comparator circuit provides an output when the reference voltage is greater than the integrated input voltage but this falls to zero when the two are equal. The output of the comparator is connected to a 'gate' circuit which opens and allows the output from a

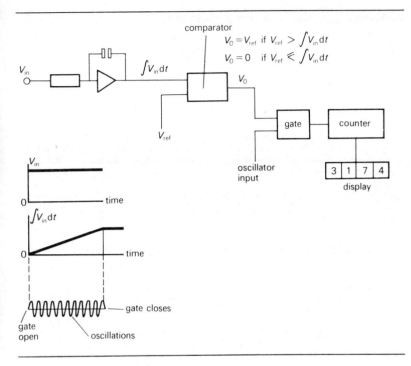

Fig. 2.32 Principle of the digital instrument.

constant frequency oscillator to be fed to a counting circuit all the while an output from the comparator is present and closes when the comparator output is zero. (This will be recognised as an AND circuit later in the book.) The number of oscillations in the time interval is directly proportional to time. (This principle occurs again in Chapter 4, Section 4.14.) The time interval is in turn directly proportional to the input voltage. The range of the instrument can be changed by using a different integrating time, i.e. by changing the time constant of the integrator circuit either by a different value of capacitor or a different value of resistor.

Summary

Pointer instruments are usually one of three types, moving coil, moving iron or dynamometer. All can be used for measuring current or voltage but the dynamometer is extensively used for power measurement. They can be used in d.c. and a.c. circuits but the moving coil meter requires a bridge rectifier and does not read r.m.s. values. The cathode ray oscilloscope can be used to measure voltage magnitude, phase difference and frequency. It can also display and compare waveforms.

Resistance, capacitance and inductance can be measured most conveniently by Wheatstone bridge networks. The potentiometer can be used for precise voltage and e.m.f. measurements. Flux measurements can be carried out by using fluxmeters, integrating circuits or Hall probes The principle of digital instruments is outlined.

Examples

1. A 50 Hz sine wave of peak value 50 V is applied to the X plates of a C.R.O. A sine wave in phase with the first wave, and of the same frequency and r.m.s. value 20 V is applied simultaneously to the Y plates. Find the angle which the trace on the screen makes with the horizontal, assuming the deflectional sensitivity of both the pairs of plates to be the same.

　　If the frequency of the second wave were 100 Hz but otherwise the conditions were unchanged, what effect would this have on the trace produced? Sketch the trace obtained.
Ans. 60.5°

2. The timebase of a C.R.O. rises linearly in 2.4 ms and the flyback time is 0.1 ms. How many complete cycles of a 2 kHz sine wave will be seen on the screen and what proportion of the last wave will be missing due to the flyback? If the sine wave frequency were changed to (*a*) 1 kHz, (*b*) 200 Hz, what effect would it have?
Ans. Four complete cycles;　20 per cent of the 5th missing

3. A resistance of 10 kΩ and a capacitor of 1 μF are connected in series to a 100 Hz, 100 V (r.m.s.) supply. The voltage across the resistor is applied to the Y plates and the voltage across the capacitor is applied to the X plates. If the Y plate sensitivity is 25 V/cm and the X plate sensitivity is 40 V/cm draw to scale the pattern produced on the C.R.O.

4. A sine wave of peak value 120 V is placed in series with an ideal rectifier (one in which the forward resistance is zero and the reverse resistance is infinite), a moving-coil ammeter, a moving-iron ammeter of negligible resistance and 50 Ω resistor. What is the indication on each meter if both have been calibrated on direct current?
Ans. 0.764 A; 1.2 A

5. If the rectifier in Problem 4 is no longer perfect but has a forward resistance of 10 Ω and a reverse resistance of 70 Ω, find the new indication on the moving-coil meter.
Ans. 0.319 A

6. A wattmeter (current coil 5 Ω, voltage coil 2000 Ω) is measuring the power in a resistance connected to a 250 V d.c. supply. If it indicates 550 W, what is the true power dissipated in the resistor assuming that the meter is connected in such a way as to produce the smaller systematic error? Show the method of connection with the terminal markings clearly indicated.
Ans. 526 W

7. A Wheatstone bridge is used to measure the resistance R_x (see Fig. 2.18). If the settings at balance of the three other arms R_1, R_2 and R_3 are 100 Ω, 10,000 Ω and 1257 Ω, respectively, what is the value of R_x?

　　If each resistance R_1, R_2, R_3 has a maximum error of ±0.2 per cent, between what limits does R_x lie?
Ans. 13.32 Ω and 11.82 Ω

8. A moving-coil meter has an F.S.D. of 10 mA and a resistance of 15 Ω. What values of resistance are needed to convert (*a*) to a voltmeter F.S.D. 10 V, (*b*) to an ammeter F.S.D. 1 A? How are these components connected to the instrument?

If the resistance in case (*b*) inadvertently changes to 0.9 x original value what is the new F.S.D. of the ammeter?
Ans. 985 Ω in series; 0.152 Ω in shunt; 1.11 A

9. A simple linear potentiometer consisting of a three-terminal resistor of total resistance 50 kΩ is placed across a 100 V d.c. supply and the moving contact is placed midway between the ends of the resistor. A 100 V meter of sensitivity 200 Ω/V is connected between one end of the potentiometer and the moving contact. What does it register?
Ans. 30.7 V

10. Draw the graphs voltage/length for the potentiometer in Problem 9, if the potentiometer winding is 20 cm long and a resistor (*a*) of 100 kΩ, (*b*) of 1 kΩ is connected between the moving contact and one end.

11. The simple potentiometer shown in the figure has a metre slidewire resistance of 0.06 Ω/cm. Find the current through the galvanometer, which has 30 Ω resistance. All other resistances may be neglected.
Ans. 15.8 mA

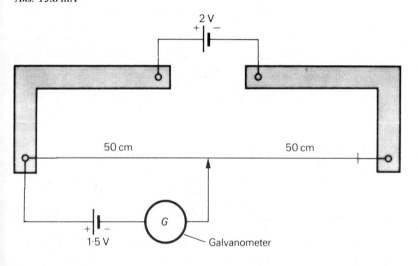

Chapter 3

Measurement of mechanical quantities

3.1 Introduction

Mechanical measurements include broadly the measurement of mass, length and angle in so far as these are basic quantities but under this general title must be included other mechanical properties such as strength, hardness and surface finish. Torque is also a mechanical quantity of interest. It is possible to make very precise mechanical measurements of length down to a 1/100 micron (10^{-8} m), for example, if sufficient precautions are taken but only rarely is such precision called for except in 'standards' laboratories and 'metrology' laboratories.

Precise mechanical measurement techniques are somewhat of a speciality but it is as well to be aware how these measurements are made.

3.2 Measurement of mass

Mass is usually measured by a comparison method. The so called 'weights' in a pair of scales are essentially masses and when using a chemical balance the mass in one scale pan is balanced against a number of known 'masses' in the other. Chemical balances can be made very sensitive and can compare masses down to 0.1 mg or better.

Larger masses are frequently measured on a spring balance. These rely in operation on Hooke's law, which states that the extension of a spring is proportional to the force exerted on it. Spring balances are essentially force-measuring devices and should be calibrated in newtons. Frequently however they are calibrated in kilogrammes and the indication on the scale is essentially the gravitational force produced on a mass. Thus on a spring balance registering 10 kilogrammes the force is 98.1 newtons. The 10 kilogrammes is often loosely referred to as the 'weight' but strictly speaking the weight of 10 kilogrammes is 98.1 newtons.

3.3 Measurement of length

Large lengths, such as the dimensions of a football pitch, are measured by a calibrated tape (Fig. 3.1), but shorter lengths can be measured accurately (down to a discrimination of 0.5 mm) using a steel rule.

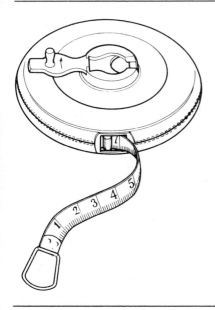

Fig. 3.1 The calibrated tape.

More accurate measurements make use of a micrometer (Fig. 3.2) or a vernier caliper (Fig. 3.3). The former relies on the principle that a small axial movement occurs when a nut is turned on a thread. One revolution

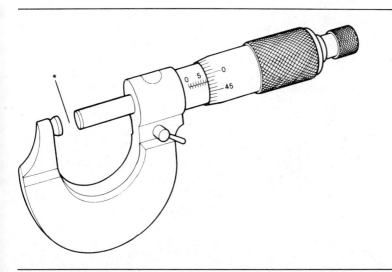

Fig. 3.2 The micrometer caliper.

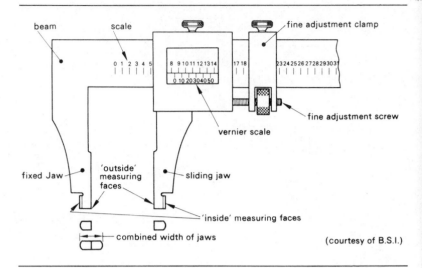

Fig. 3.3   The vernier caliper.

of the nut produces a movement along the axis of one pitch of the thread.

The metric micrometer has a pitch of 0.5 mm. One complete revolution of the thimble closes (or opens) the gap between the faces (marked * in the diagram) by 0.5 mm. The thimble is divided into fifty divisions and each division therefore corresponds to 0.01 mm.

The reading on the micrometer is given by: (1) the number of 'whole' millimetres marked on the upper half of the scale on the barrel, plus (2) the number of half millimetres marked on the lower half of the scale, plus (3) the reading on the thimble markings coincident with the datum line. The micrometer in Fig. 3.4 reads 9.98 mm.

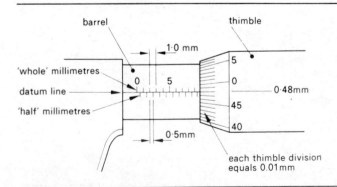

Fig. 3.4   Micrometer scales (metric).

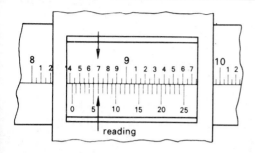

Fig. 3.5 A 25-division vernier scale.

The vernier caliper uses the fact that the vernier scale has one more division than the corresponding length on the main scale.

The first two significant figures can be read off the main scale opposite the figure 0 on the vernier. Thus in the enlarged figure shown (Fig. 3.5) the first two figures are 8.4. The next significant figures occur where the main scale divisions and the vernier divisions are coincident — six in this case. The twenty-five divisions on the vernier correspond to a half division on the main scale. The reading of the vernier scale must therefore be multiplied by 2. So the next figures must be 6 x 2 = 12. The reading is therefore 8.412.

Dial gauges provide a third method of precise measurement of length (Fig. 3.6). A mechanical amplification of the movement of the plunger is achieved by a form of rack and pinion arrangement. The zero of the instrument can be adjusted by movement of the bezel. Dial gauges measure essentially changes of length rather than absolute distances.

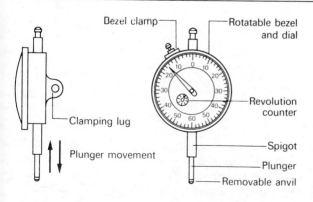

Fig. 3.6 Dial gauge.

All the instruments described have the same order of accuracy but the micrometer and dial gauge give rather better discrimination — down to 0.01 mm.

Micrometers, vernier calipers and dial gauges can have their accuracy checked fairly easily by the use of 'slip gauges' (see p. 5).

There are many variations of the micrometer and the vernier caliper which allow both inside and outside measurements to be made as well as depth.

3.4 Measurement of angle

Protractors are used for measuring angles. The more sophisticated version employs a vernier scale. The application of the device is largely self-explanatory. The calibration of a protractor or the setting up of a precise angle can be achieved by using a sine bar (Fig. 3.7). The distance L is known precisely (200 mm in this case).

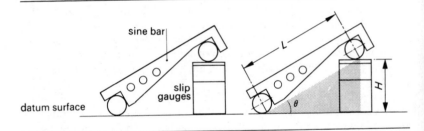

Fig. 3.7   Principle of the sine-bar.

If an angle of 25° is needed the slip gauges giving height H are calibrated simply from

$$\sin \theta = \frac{H}{L} \quad \text{hence } H = L \sin 25°$$
$$= 200 \times 0.4226$$
$$= 84.52 \text{ mm}$$

3.5 Tensile testing

One of the important tests applied to materials testing is the behaviour of a particular material when a force is applied which stretches it. For most materials used in engineering the amount of 'stretch' is small, so that some means of measuring a small extension is necessary. A number of alternatives exist:

1. A dial gauge can be employed. Although its sensitivity is perfectly

adequate this method is seldom used as the dial gauge could be easily disturbed as the load (force) is applied to the specimen under test.

2. A strain gauge may be used. This changes the mechanical movement into a change of electrical resistance. Chapter 7 deals in greater detail with this device.

3. An extensometer may be employed. This is an instrument which is clamped rigidly to the test specimen and a mark on the specimen is aligned optically with cross wires in the eye piece. As the material is stretched the cross wires are displaced from the original mark. The wires are realigned by means of a mechanical linkage (Fig. 3.8) and the amount of realignment is recorded on the micrometer screw.

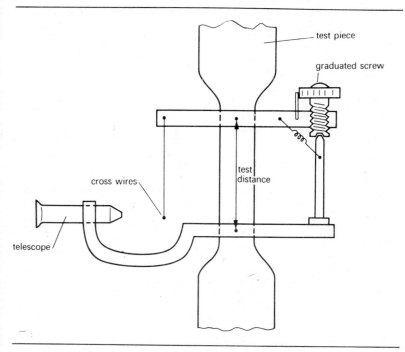

Fig. 3.8 Ewing extensometer.

The first and third method suffer from the fact that the measurement has to be made *in situ*. The strain gauge enables measurements to be recorded some distance from the testing apparatus.

The Hounsfield tensometer is a bench-type testing apparatus which enables the load and extension of the test specimen to be recorded automatically.

The graph obtained plots load to a base of extension and is illustrated in Fig. 3.9. The section *ab* is linear and up to this point the material is elastic and obeys Hooke's law.

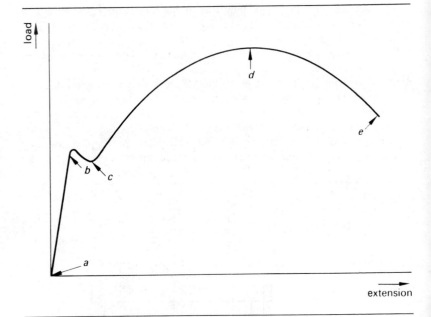

Fig. 3.9   Load/extension graph — mild steel (annealed).

Beyond point *c* the material behaves in a 'plastic' manner and will not return to its original length if the load is removed.

d is the ultimate stress point and *e* is the breaking point. The two quantities which are derived from this type of test are

$$(i)\ \text{stress} = \frac{\text{Load}}{\text{original cross sectional area}}$$

stress can be measured in newtons/square metre

$$\text{and (ii)}\ \text{strain} = \frac{\text{extension}}{\text{original length}}$$

strain is non-dimensional.

Young's modulus *E* is the ratio stress/strain and has the same dimensions as stress.

3.6 Measurement of hardness

A number of different pieces of apparatus are commercially available for the measurement of hardness of a material but all use the same principle. A hard steel ball or a shaped diamond wedge is pushed on to the surface of

improved
loading indicator

wide range of
loading speeds

finger-tip control for
loading and unloading

ample room
for specimens

built-in power unit

shockless loading by high
precision lever and weights

Fig. 3.10 Brinell testing machine

the material being tested with a known force. The depth of penetration is measured using a microscope and this is then a measure of the hardness. The Brinell machine is shown in Fig. 3.10 and the depth of penetration

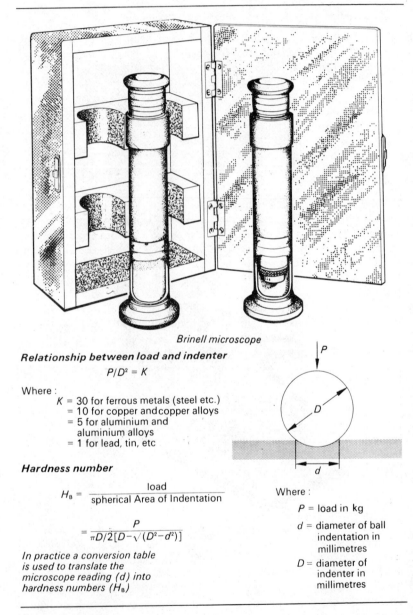

Brinell microscope

Relationship between load and indenter

$$P/D^2 = K$$

Where :

    K = 30 for ferrous metals (steel etc.)
      = 10 for copper and copper alloys
      = 5 for aluminium and
        aluminium alloys
      = 1 for lead, tin, etc

Hardness number

$$H_B = \frac{\text{load}}{\text{spherical Area of Indentation}}$$

$$= \frac{P}{\pi D/2 [D - \sqrt{(D^2 - d^2)}]}$$

In practice a conversion table is used to translate the microscope reading (d) into hardness numbers (H_B)

Where :

P = load in kg

d = diameter of ball
    indentation in
    millimetres

D = diameter of
    indenter in
    millimetres

Fig. 3.11   Brinell test.

can be calculated from the measurement of the diameter of the circle made by the ball in the material (Fig. 3.11).

The hardness number is given by

$$H_B = \frac{\text{Load}}{\text{Spherical area of indentation}}$$

Although this can be calculated from first principles the commercial machine has a conversion table for changing the microscope reading into the hardness number.

The Brinell scale extends up to 10,000, cutting steels having a hardness around 1000 and plastics in the range 10—50. It is an arbitrary scale and does not compare in numerical range with hardness numbers obtained from other testing machines, e.g. Rockwell.

3.7 Surface finish

A material such as steel which has been through a machining process (cutting, filing, shaping, grinding, etc.) shows some irregularity of surface. The irregularity may be caused by the machine tools but is also a feature of the material itself. Polishing and 'lapping' will remove the worst of the irregularities but even then some 'hills and dales' will remain. One quantitative method of measurement of surface finish is to move a stylus across the surface and then to amplify any movement produced. This is most conveniently carried out electronically and the movement is then recorded. In the example given both axes are magnified although the vertical axis is to a much greater extent (Fig. 3.12). It is then possible to produce a number which gives a quantitative measure of the smoothness of the surface. The Taylor Hobson 'Talysurf' actually uses the signal from a detecting head to modulate a carrier (see p. 152). The action of the detecting head is somewhat similar to the pick-up of a gramophone but in this case the two windings form two arms of a balanced bridge network (Fig. 3.13). Any movement of the stylus unbalances the bridge and an output at the oscillator frequency is applied to the amplifier. The amplitude of the output varies with the movement of the stylus and hence the surface finish being measured. This is amplitude modulation. After amplification the carrier is removed by a demodulation process and the signal is then recorded.

A different approach to the measurement of surface finish uses the interferometry principle.

When the rays of light from a monochromatic source travel along different paths a phase difference is likely to be present. Where the phase difference is 180° between one ray and another cancellation occurs and the light is reduced in intensity (possibly cut off completely). Where the phase difference is zero (or 360°) reinforcement occurs and the light is increased in intensity. A surface illuminated in this way shows a series of

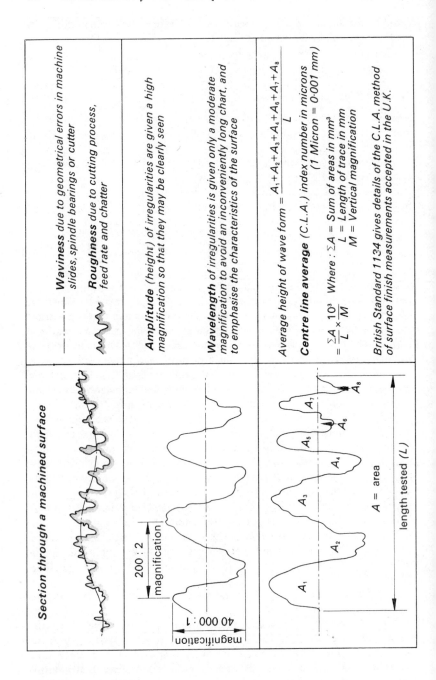

Section through a machined surface	
	— · — · — **Waviness** *due to geometrical errors in machine slides, spindle bearings or cutter*
	⋀⋁⋀ **Roughness** *due to cutting process, feed rate and chatter*
200 : 2 magnification 40 000 : 1 magnification	**Amplitude** *(height) of irregularities are given a high magnification so that they may be clearly seen*
	Wavelength *of irregularities is given only a moderate magnification to avoid an inconveniently long chart, and to emphasise the characteristics of the surface*
$A = $ area length tested (L)	*Average height of wave form* $= \dfrac{A_1 + A_2 + A_3 + A_4 + A_5 + A_6 + A_7 + A_8}{L}$ **Centre line average** *(C.L.A.) index number in microns* *(1 Micron = 0·001 mm)* $= \dfrac{\Sigma A}{L} \times \dfrac{10^3}{M}$ *Where :* $\Sigma A = $ *Sum of areas in* mm^3 $L = $ *Length of trace in mm* $M = $ *Vertical magnification* *British Standard 1134 gives details of the C.L.A. method of surface finish measurements accepted in the U.K.*

Fig. 3.12   Surface finish.

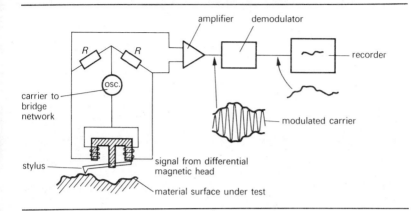

Fig. 3.13   Principle of 'Talysurf'.

light and dark areas where either cancellation or reinforcement of the light rays take place. If a section of optically flat glass (glass which is highly polished with opposite faces exactly parallel) is placed above a reflecting steel block, rays will be reflected from the inner surface of the glass and also from the surface of the steel. Depending upon the different lengths of path taken by the light rays, which in turn depend upon the angle between the steel and the glass, so the light will be reinforced or cancelled out.

Viewed through the glass the surface will have a series of dark parallel stripes if the steel has a polished surface comparable in flatness and surface finish with that of the glass. Any variation in surface finish of the steel will cause the reflected light rays to take a different path and hence change the 'interference' pattern.

Slight concavity of the surface, irregularities or scratches clearly show up by this method although it is rather difficult to put a total quantitative assessment on the results. It is however possible to estimate the depth of a scratch, for example, since the spacing of the fringes correspond to the half wavelength of the light source which is known to a high degree of precision.

3.8 Torque measurements

Static torque measurement poses no problems of measurement. One can either apply or measure a torque on a shaft by clamping an arm of known length to the shaft. The end of the shaft is loaded by means of a spring balance or a gravitational load. The torque is then given by $d/2$ (the distance from the shaft to the point of load application) x the force exerted in newtons. If a gravitational load is used remember that the force is mass x g (9.81 m/s^2).

If the shaft is rotating then normally two methods are available: (i) the

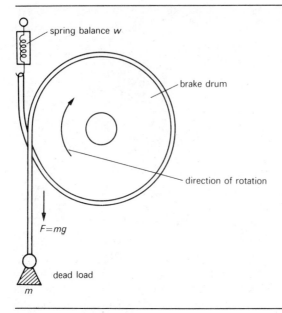

spring balance *w*

brake drum

direction of rotation

F=mg

dead load

m

Fig. 3.14   The Prony brake.

Prony brake and (ii) the Dynamometer brake. The former is shown in Fig. 3.14. The dead load is being lifted by the frictional force of the rope on the brake drum, the rope acting essentially as a brake band. The effective force exerted on the perimeter of the drum is the difference between the dead load and the spring balance reading, i.e. $mg - w$ (w must be in newtons). The torque is then given by

$$\text{Torque T in newton metres} = (mg - w) \times \frac{d}{2}$$

where d is the brake drum diameter in metres. To be a little more precise, the effective diameter is that of the drum plus the thickness of the rope (or brake band).

Considerable heat is often generated in the process (the output from the rotating shaft is $2\pi nT$ watts if n is the rotation speed in revolutions per second) and it is necessary to cool the brake drum, usually by water.

The dynamometer is essentially a d.c. shunt generator whose field windings are arranged on trunnions so that they can move through a small angle ($20°$ approximately). The armature is fixed to the shaft which in turn is attached to the machine which is under test. When the shaft rotates the generator produces a terminal voltage which is applied to a resistive load. At a given speed, altering the value of the resistance changes the power generated and affects the torque on the shaft. This produces a reaction on the field winding which causes it to swing through an angle. (It is prevented from moving more than the 20° stated by mechanical stops.) A mechanical load is now applied which will bring back the swinging field to its original position. The torque then applied is the force x length of torque arm (Fig. 3.15).

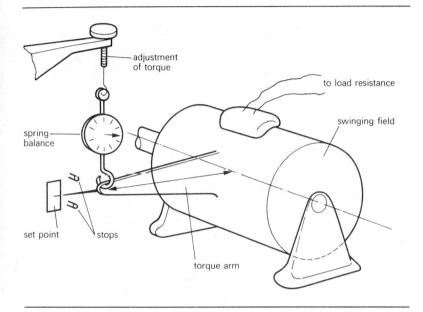

Fig. 3.15   Swinging field dynamometer.

While the second method requires more sophisticated equipment the torque can be applied and measured rather more easily and accurately. An electrical transducer can be employed as a third alternative (see Chapter 7).

Summary

Mechanical measurements cover the fields of length, mass, angle, strength, surface finish, hardness and torque. Some mechanical measurements can be extremely precise.

Two common length-measuring devices are the micrometer and the vernier caliper.

Interferometry provides a sensitive means of noting surface imperfections.

Torque measurements can be made by a mechanical brake or by a 'swinging' field dynamometer.

Questions

1. Are there means of measuring length to an accuracy better than 0.01 mm other than those described in this chapter? If so, give details.

2. What are the readings on the two calipers shown?

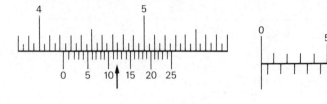

Chapter 4

Measurement of other quantities

4.1 Introduction

The division between mechanical, electrical and 'other' quantities is very arbitrary, as many measurements involve some electrical or mechanical measurement. A few however (many of these in the chemical field) rely on something rather subjective such as a colour change. The quantities in this chapter are essentially those not already covered which do not obviously fall in the categories of mechanical and electrical measurements. The first 'other quantity' will be in the area of heat measurements.

Temperature

The SI unit of temperature is the **degree Kelvin**. The unit employed frequently in industry is the degree Centigrade or degree Celsius. The relationship has already been mentioned (p. 4).

Temperature is measured in three ways:

1. A mechanical change produced by the temperature, e.g. the expansion of a liquid.
2. An electrical change – this is the thermoelectric effect.
3. A physical or chemical change, e.g. the melting of a wax.

4.2 Temperature measurement by mechanical means

The first of these alternatives is exemplified by the mercury thermometer. Here an expansion in the mercury caused by a temperature change forces a column of mercury along a calibrated tube. The accuracy depends upon the care with which the thermometer is calibrated in the first instance and the uniformity of the bore of the tube containing the mercury. The discrimination (or sensitivity) increases as the cross section of the bore decreases but for a given range this increases the length of thermometer. If only a relatively narrow range of temperature measurement is required about some fixed value it is possible to produce thermometers with a varying tube area. High sensitivity (or discrimination) is then possible. The clinical thermometer is such an example where temperatures varying only slightly about normal body temperature (98.4°F or 36.9°C) need to be measured. The clinical thermometer has a sensitivity such that a change of

0.1°C can be easily seen. The temperature range possible with the mercury thermometer is limited by the freezing point and the boiling point of mercury. Normally this type is seldom employed above 300°C. The tube containing the mercury is often glass (in order that the column of liquid can be seen) but this means that the thermometer cannot be used for giving readings distant from the point at which the measurement is made. A variation in design — the mercury in steel thermometers — enables distant measurements to be made and provides a more robust type of measuring device (Fig. 4.1). This uses the pressure generated by mercury vapour and can be used at higher temperatures.

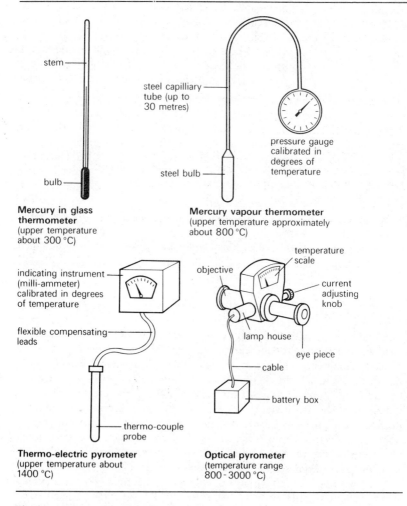

Fig. 4.1   Temperature-measuring devices.

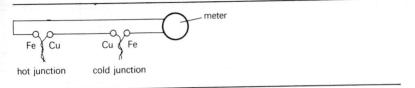

Fig. 4.2   Thermocouple.

4.3   Temperature measurements using electrical effects

The more common industrial temperature indicator employs a thermo-couple which is capable of being used at much higher temperatures (up to 1450°C). The principle of the thermocouple is shown in Fig. 4.2 and relies on the fact that the junction of two dissimilar metals when heated produces an electron movement across the junction. When a difference of temperature exists between the cold junction and the hot junction an e.m.f. is generated which causes current to flow through the circuit. The indication on the meter is a measure of the temperature. For temperatures up to 750°C the metals employed are usually iron and constantan (a nickel copper alloy). This type of junction produces an e.m.f. of 0.054 mV for each degree change in temperature and this means that the 'sensitivity' is not high. For the highest temperatures (up to 1450°C) a platinum/platinum–rhodium junction is used but the sensitivity is even lower (0.01 mV/°C approximately). A potentiometer may be employed to measure more accurately the e.m.f. If this potentiometer is of the self-balancing type then it is possible to record the temperature by a suitable pen-recording mechanism. If the temperature range required does not extend beyond about 600°C then a resistance thermometer is often employed (Fig. 4.3). This device relies on the fact that many materials change their resistance with changes in temperature. The resistance thermometer element forms one arm of a Wheatstone bridge and changes in resistance caused by temperature variations unbalances the bridge. By

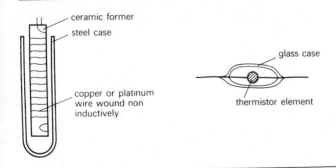

Fig. 4.3   Resistance thermometer.

rebalancing the bridge the resistance change (and hence the temperature change) can be measured. Again if the bridge is self balancing the temperature can be easily recorded. Platinum, nickel or copper are usually chosen for the resistance element.

Both the thermocouple and the resistance thermometer can be used some distance from the point at which the temperature is indicated but it is necessary to take into account any effect of the connecting leads which might contribute to the e.m.f. generated or the resistance of the variable element (see Fig. 4.4).

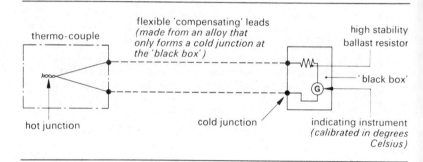

Fig. 4.4   The thermocouple pyrometer circuit.

Sometimes when the temperature to be measured is relatively low a cintered mixture of certain metal oxides is used as the sensing element. This has two advantages: (i) the percentage change in resistance/°C is greater than that of a metal and (ii) since it can be made much smaller (Fig. 4.3) the temperature at a given spot can be more easily measured. Such a device is usually called a thermistor and its small size makes it attractive in the measurements of power in microwaves. It has incidentally a negative temperature coefficient.

4.4 Methods of temperature measurement using chemical or physical changes

Many materials change either their physical properties or their chemical properties with increase in temperature. Carbon steel for example changes from a pale straw colour at about 220°C to a deep blue at about 300°C. Special heat-sensitive paints or coloured waxes are available which change colour or melt at specified temperatures. These are employed in certain heat treatment processes where it is necessary to know when a particular temperature has been reached at a given point. These methods of temperature measurement cannot be regarded as particularly accurate as there is no reliable means of knowing to what extent a temperature has been exceeded.

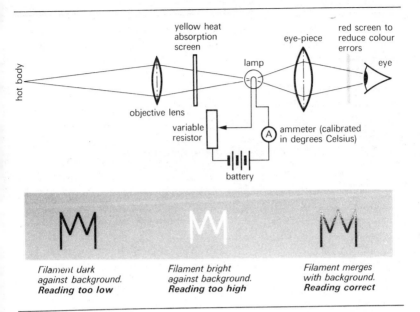

Fig. 4.5   The optical pyrometer.

The optical pyrometer (Fig. 4.5) is a much better method of measuring high temperatures (above about 600°C). The temperature of the heated body to be measured causes light emission (from dull red to white) and the current through the filament is adjusted until it disappears when viewed through the eye piece. At this point the temperature of the filament is equal to the temperature of the object being heated. The current is read on a meter which is calibrated in degrees Celsius.

4.5   Errors in temperature measurement

Other than calibration errors the main source of error in temperature measurement lies in the personal observational error — especially in the case of methods involving colour change. The systematic error where the introduction of the temperature-sensing device disturbs the system being measured is also sometimes of significance. This latter point becomes increasingly important when the physical size of the thermometer is a significant proportion of the system itself. A small cup of hot water may easily have its temperature changed by the introduction of a cold thermometer. A bath full of hot water is unlikely to be affected in the same way. (The optical pyrometer does not introduce this sort of error but cannot be used at relatively low temperatures.) There is in addition a delay in indication when the sensing device is actually placed in contact with the process whose temperature is being measured. It may take seconds or even minutes before the sensing element reaches the temperature of the quantity being measured. A clinical thermometer for example requires at least 30 seconds before it reaches body temperature.

It is therefore difficult to record temperatures which fluctuate rapidly due to this time lag. It is necessary then to use sensing elements which are physically small and have therefore a small thermal capacity.

4.6 Measurement of the calorific value of a fuel

The calorific value of a fuel is the total heat given out per unit mass of the fuel when combustion is complete. The principle of measurement whether gas, liquid or solid fuel is to be measured is the same but the actual apparatus differs appreciably in the three cases. Basically one has to weigh the quantity of fuel, ensure that combustion is complete and measure the total amount of heat produced. Only one method is described here and that is the one employed with solid fuel.

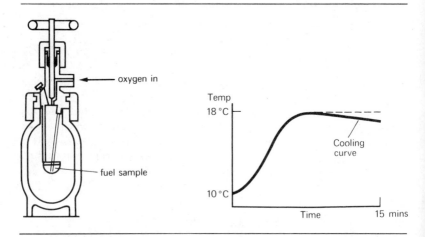

Fig. 4.6   The bomb calorimeter.

A device called the Bomb Calorimeter is used (Fig. 4.6). A known quantity of fuel in a powdered form is placed in the receptacle. Oxygen at a considerable pressure (about 20 atmospheres) is introduced and the inlet is then sealed off by operating the control valve. The whole calorimeter is immersed in a known quantity of water whose temperature has been measured. The fuel is ignited by passing a current through a heating coil near or actually in the receptacle. The temperature of the water is monitored for some minutes after the initial ignition.

The total heat output is (the mass of water + water equivalent of the bomb calorimeter) x rise in temperature.

The combustion may take several minutes and a typical temperature/ time graph is also shown in Fig. 4.6. Allowance can be made for temperature loss during combustion by noting the drop in temperature per minute after combustion is complete.

A thermometer with high discrimination is usually needed because the temperature rise of the water is small.

4.7 Light output measurements — definitions

Light has the properties of brightness (or luminance) and colour and when making measurements on the light output from different forms of lamps the colour content of the light source can affect the result. Light is radiated energy which can be seen with the human eye and therefore covers a spectrum of colour. The response of the human eye is shown in Fig. 4.7. White light is a mixture of light of different wavelengths.

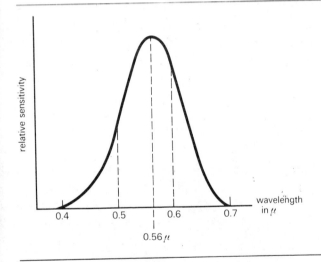

Fig. 4.7   Response of human eye.

Luminous flux is the light energy emitted per second and is therefore a unit of power. Luminous intensity is the luminous flux radiated per unit solid angle, that is it is a directional light power.

The unit of light is based on the **candela**, which is the luminous intensity of each 1/60 of a square centimetre of surface of the metal platinum at its temperature of solidification (2042 K).

The rather unusual definition is an attempt to make the present-day primary standard not appreciably different from the old 'candle power' unit.

The **lumen** is the luminous flux emitted per unit solid angle from a point source of 1 candela having uniform intensity.

It follows that the total luminous flux from a source of 1 candela is 4π lumens.

Illumination is the flux received per unit area. The present-day unit is the **lux** or lumen per square metre.

4.8 Measurement of luminous intensity

The normal measurement employs essentially a comparison method. Two lamps A and B are arranged either side of a screen and the distances x_1 and x_2 are adjusted until the illumination appears the same on either side (Fig. 4.8)

The $\dfrac{\text{luminous intensity of A}}{x_1^2} = \dfrac{\text{luminous intensity of B}}{x_2^2}$

since the inverse square law is obeyed.

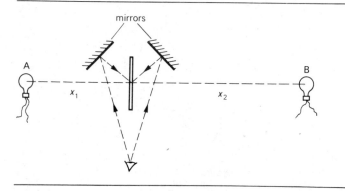

Fig. 4.8   Comparison of illumination.

A number of different methods exist of illumination comparison, the simplest being the 'grease spot' photometer and a more sophisticated form being the flicker photometer whereby, by an arrangement of mirrors, the illumination from the two surfaces is presented to the eye alternately. The speed of alternation is made variable and the optimum conditions depend upon the colour content of the two sources of light. When the illumination from the two sources is the same the flicker disappears. With the grease spot, equal illumination from either side causes the spot to disappear.

In general the luminous intensity from a lamp is not uniform and varies in both the vertical plane and the horizontal plane. A filament lamp for example exhibits zero intensity immediately about the lamp holder. Polar diagrams can be drawn showing the variation in intensity (Fig. 4.9).

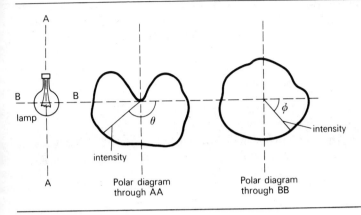

Fig. 4.9   Polar diagram.

A quantity often required by lamp manufacturers is the average intensity in all directions. This is termed the Mean Spherical Intensity and can be obtained from the vertical polar diagram if certain assumptions are made. In practice the measurement of the mean spherical intensity can be best carried out using an integrating sphere (Fig. 4.10). It can be shown that the illumination produced at any point on the inside surface is the same. The total light flux is responsible for this illumination and therefore the brightness at a small window is a measure of the total light flux and hence the mean spherical intensity. For accurate results only reflected

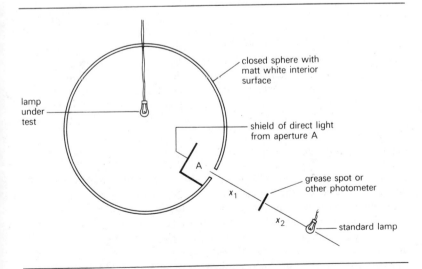

Fig. 4.10   Integrating sphere for measuring mean spherical intensity.

light must contribute to the measurement. Usually a comparison is made with a light source of known intensity, the measurements at the window being made in turn with the unknown and then the standard source.

4.9 The photo voltaic cell

A device called the photo voltaic cell, similar to the light meter used by photographers, is frequently employed in light intensity measurements. Its construction is shown in Fig. 4.11 and its action relies on the formation of

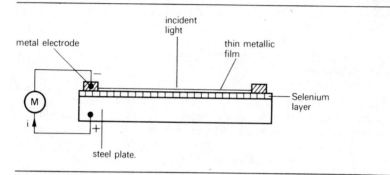

Fig. 4.11 Photo voltaic cell.

a semiconductor rectifier junction between the top and bottom electrode. When the top is illuminated electrons cross the rectifier junction and produce an e.m.f. which in turn causes current to flow if the external circuit is closed. A sensitive meter records this current which is proportional to the light flux falling on the cell. The response of photo voltaic cells is different with differently coloured light sources and does not correspond very closely to the spectral response of the human eye. However, colour filters can correct this.

The photo voltaic cell can be used with an integrating sphere in place of the photometer arrangement shown in Fig. 4.10.

4.10 Sound measurements

Sound is a subjective phenomena in much the same way as light is subjective. In the latter it is the human eye which perceives light, in the former it is the human ear which hears sound. Both light and sound are transmitted by waves but with light the waves are transverse (i.e. at right angles to the direction of propagation). With sound the waves are longitudinal (i.e. in the same plane as the direction of propagation).

Light can be transmitted through a vacuum but sound requires a medium (solid, liquid or gas) to provide the means of transmission.

Sound is characterised by pitch — or frequency — and loudness. (These correspond to colour and brightness with light. The pitch of a note is the subjective measurement of frequency in the audio range. While the two are very close it can be shown that the sound level can affect the subjective judgement of frequency.)

Sound is energy and the rate at which sound is emitted is measured non-subjectively in joules per second or watts. The sound passing through a given area is the sound intensity and is measured in watts per square metre.

The human ear does not hear sound intensities of different frequencies uniformly and is normally more sensitive to sounds in the frequency range 1–5 kHz. Few adults hear frequencies above about 17 kHz but it is believed that babies can hear much higher frequencies. The lowest intensity which can be heard is about 10^{-12} W/m². At intensities above 1 W/m² pain is experienced and permanent damage may be caused to the hearing mechanism of the brain. Some animals (bats for example) have a much higher hearing frequency range than humans. The response of the average human ear is shown in Fig. 4.12, plotting sound intensity against frequency. Quantitative measurements of sound intensity employ a logarithmic scale in decibels (dB) defined as follows:

The power ratio a to b in decibels (dB) is $10 \log_{10} \dfrac{\text{power } a \text{ in watts}}{\text{power } b \text{ in watts}}$

(This ratio occurs again in Chapter 8.)

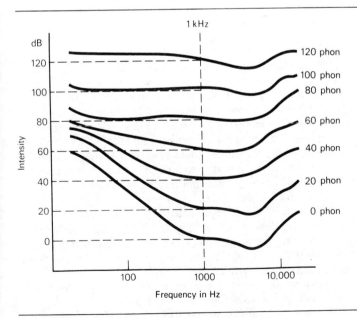

Fig. 4.12 Frequency response of the human ear.

The zero of the sound intensity scale is arbitrarily taken at a frequency of 1 kHz at a value of 10^{-12} W/m^2, which is the limit of audibility for most people. Logarithmic subjective measurements above this are measured in **phons**. Thus 0 phons is the limit of audibility and varies with frequency. One hundred and twenty phons is the threshold of pain and corresponds to 1 W/m^2 at 1 kHz.

The non-uniform frequency response of the human ear makes sound measurements difficult as sound levels which are barely audible at one frequency are uncomfortably loud at another. The sound level meter should ideally have sound filters fitted so that the sensitivity of the meter corresponds to the average human ear if subjective measurements are to be made. If the sound level meter measures sound intensities uniformly over the audible frequency range (nominally 20 Hz–20 kHz) the measurement is recorded in dB(C). If the sound level meter is modified so that its response matches reasonably well that of the human ear the measurement is recorded in dB(A). It is the latter measurement which is used in acoustic measurements of noisy areas, e.g. airfields.

4.11 Sound in buildings

Generally speaking most working areas require unwanted sound (noise) to be kept down to a minimum. High noise levels can produce a stress situation in human beings so that it is desirable to reduce noise as far as possible. Unwanted sound can sometimes be absorbed by using a form of silencer at the noise source (e.g. a car silencer) or absorbent material (e.g. carpeting or polystyrene) near the source.

Excessive absorption of sound can equally well produce undesirable effects. A room in which as much as possible of the sound generated within it is absorbed is called an anechoic chamber. Working in such conditions for any length of time can also produce unpleasant effects. ('The silence was painful.') Sound intensity at any point decays in an exponential manner with time and the reverberation is a measure of the rate of decay.

Reverberation time is the time for the intensity to fall 60 dB, i.e. to 10^{-6} of its initial value. Reverberation time should be small for offices – about ½ second, but for concert halls figures of up to four times this value are normal.

Long reverberation times give a hollow, empty sound and short reverberation times give a dead 'closed-in' effect. The design of concert halls for good acoustical properties is extremely difficult, because it is often the architect who designs the building and the acoustic engineer has to do his best with the hall given him. Good concert halls often have a floor which is isolated from the main building in order that sound from outside is not transmitted to the hall.

Sometimes acoustic deficiencies can be made good by introducing baffles or absorbent plates (this has been done at the Albert Hall) or artificially modifying the acoustics by amplifying certain frequency ranges (as has happened in the Festival Hall).

'Good' acoustical properties are still very much a matter of opinion and the good acoustic properties of some cathedrals rely as much on the dimensions of the nave and transept as anything. It is unlikely that there was any attempt at acoustic engineering in medieval times.

4.11 Sound level instruments

The basis on which many sound measurements are made rely on the variation in air pressure. Such variations cause the diaphragm of a microphone (or loudspeaker) to move which in turn causes a coil to move in a magnetic flux (see Fig. 4.13). This movement produces an e.m.f., which when

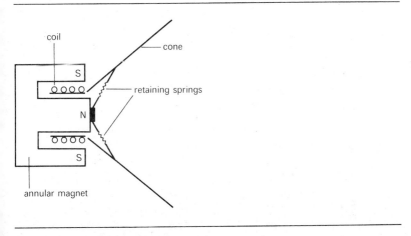

Fig. 4.13 Moving coil loudspeaker.

suitably amplified is applied to a meter. The meter indication is a measure of the sound intensity reaching the diaphragm. It is possible to modify the response by introducing filtering networks. There is no standard unit of sound so that sound measurements are essentially comparisons.

It is possible to use a piston in a closed cylinder to produce a known air pressure and hence by moving the piston up and down to produce a variation in air pressure at a given frequency. A microphone placed at the end of the cylinder will then receive a known amount of power at an audio frequency. If the microphone forms part of a sound level meter a calibration point can then be obtained. This method of calibration is useful only at relatively low frequencies. At higher frequencies the air pressure variation is produced by the movement of a plate of a capacitor caused by the application of a sine wave of voltage. A measurement of the change in energy as the capacitor plate is moved is related to the sound output power. It is difficult however to make accurate sound measurements.

4.12 Fluid measurements — pressure

There are two basic measurements in the field of fluids. One is concerned with pressure and the other with flow. The simplest pressure indicator is the manometer (Fig. 4.14). The pressure in the vessel causes the mercury

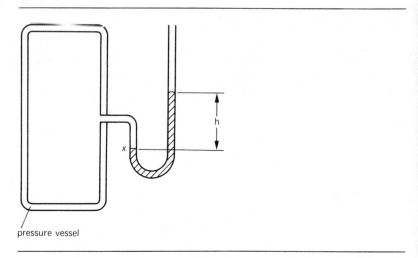

pressure vessel

Fig. 4.14   The simple manometer.

in the tube to be displaced. If h is the difference in level between the columns of mercury it follows that the force at point 'x' due to the mercury is $h \times a \times \rho \times g$ where h is the height, a is the cross section of the tube, ρ is the density of the liquid and g the gravitational constant.

The pressure at the interface of the mercury and the gas or liquid in the vessel is this force ÷ the cross sectional area of the tube

i.e. pressure $= \dfrac{ha\rho g}{a}$

Hence the pressure $= h\rho g$ Newtons per square metre

if h is in metres

ρ is in kilo/m^3

g is in m/s^2 (9.81 m/s^2). This result is independent of the cross section 'a'.

The total pressure at the interface is due to the atmospheric pressure acting as well. Hence pressure $P_{\text{total}} = h\rho g + P_A$ where P_A is the atmospheric pressure in N/m^2.

The sensitivity of this device can be increased by using a liquid other than mercury in the manometer tube which has a lower density. Water

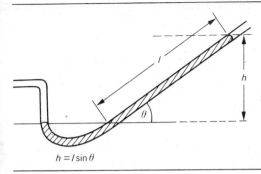

$h = l \sin \theta$

Fig. 4.15   Sloping tube.

(coloured by a suitable dye so that it can be more easily seen) is sometimes used.

The inclination of the calibrated tube can be moved from the vertical to space out the indications. The effective height of the liquid is then $l \sin \theta$ (Fig. 4.15). This also increases the sensitivity. The difference in height of the liquid in the two sides of the U tube is somewhat annoying as two, rather than one, measurements are required. This problem can be overcome by having a large vessel one side of the 'U'. The level in this vessel does not change appreciably with movement in the level in the calibrated tube since the volume of liquid displaced is small (Fig. 4.16).

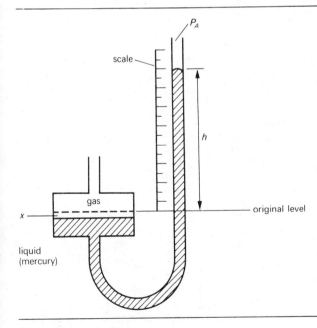

Fig. 4.16   Modified manometer.

The pressure at the interface of the gas and liquid is P_1, hence the pressure due to the displacement of the liquid in the manometer tube must equal this:

i.e. $P_1 = \rho(b + x)g + P_A$

But the volume of liquid displaced in the large vessel is Ax where A is the area of cross sectional.

Thus $Ax = ab$

or $x = \dfrac{ab}{A}$

Hence $P_1 = \rho \left(b + \dfrac{ab}{A} \right) g + P_A$

or $P_1 = \rho bg \left(1 + \dfrac{a}{A} \right) + P_A$

if $a/A \ll 1$ then $P_1 = \rho bg + P_A$. (Note that b is the change in level from the original not the difference in level between the liquid in the large vessel and the calibrated tube.)

An alternative method of pressure measurement employs the Bourdon tube (Fig. 4.17). The pressure causes the flexible tube to straighten and the movement of the end of the tube is amplified mechanically by the use of a rack and pinion device. As the tube is closed at one end there is no additional atmospheric pressure P_A to be taken into account.

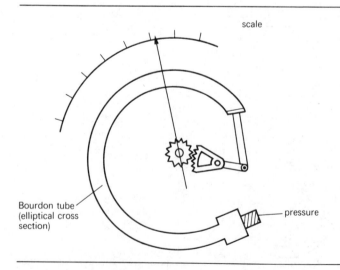

scale

Bourdon tube
(elliptical cross
section)

pressure

Fig. 4.17   Bourdon pressure gauge.

The calibration of pressure gauges is relatively easy using the displacement of a mass with pressure or alternatively using a manometer of the type already described. The calibration of the manometer depends only upon the density of the displaced liquid, the height measured and the gravitational constant all of which can be measured accurately.

Rapid changes of pressure are more easily measured by using a transducer (see Chapter 7).

4.13 Fluid measurements — flow

The quantity of a liquid or gas passing a given point per second is the flow of a liquid or more strictly the volume flow rate $\dfrac{dQ}{dt}$. It is essentially the velocity of the liquid or gas.

The total quantity of liquid or gas delivered in an interval δt is given by $\dfrac{dQ}{dt} \cdot \delta t$, i.e. the total quantity of liquid or gas delivered is the summation or integration of the flow rate $\left(\dfrac{dQ}{dt} \right)$ over a given interval of time.

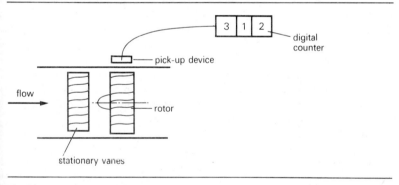

Fig. 4.18   Rotor type flowmeter.

Two basic approaches to the problem are available, one is to make the flow drive some form of rotor. The speed of rotation of the rotor is then an indication of the flow rate. This is illustrated in Fig. 4.18. The summation of the rotation, i.e. counting the number of revolutions, gives the total quantity. The summation can be carried out by mechanical means by a normal counting mechanism as employed in a car speedometer. Here the speed is indicated directly and the mileage meter (odometer) is a summation of the speed. If the speed or flow rate is proportional to an electrical voltage then the summation can be carried out using an integrating circuit. When the quantity of liquid is relatively large a venturi meter can be

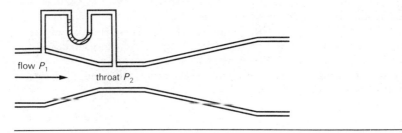

Fig. 4.19 Venturi meter.

employed (Fig. 4.19). The principle of operation depends upon the application of Bernoulli's equation. The pressure at the 'throat' is lower than the pressure at the approach side.

It can be shown that the flow rate is directly proportional to the square root of the pressure difference, or:

$$\frac{dQ}{dt} = k \ \sqrt{P_1 - P_2}$$

The constant k depends upon the density of the liquid whose flow rate is being measured, the areas of cross section at the throat and the main duct and the Reynolds* number. Venturi tubes should be mounted horizontally in a long straight section of piping to obtain accurate readings of flow rate.

A pitot tube (Fig. 4.20) may be used to measure air flow in a pipe. The flow rate (or air velocity) is given by a similar relationship to that in the venturi meter. The differential pressure P_1-P_2 depends upon the position of the tube in the air flow. With lower Reynolds numbers the air flow is laminar and a parabolic variation of pressure across the air flow exists. With higher Reynolds numbers the flow becomes turbulent and the variation in pressure across the flow is less over much of the diameter of the duct carrying the flow.

4.14 Measurement of time

Time can be measured very accurately indeed. The SI unit of time is the **second**, which is quoted to ten significant figures (p. 3) and this implies

---

* Reynolds number for a circular pipe = $\dfrac{v\rho d}{n}$

where v = velocity

    ρ = fluid density

    d = pipe diameter

    n = viscosity of the fluid

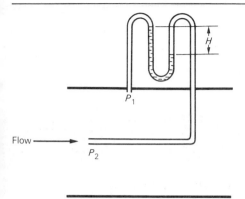

Fig. 4.20 The pitot tube.

an error of no more than a second every 30 years. Rarely have we a need to make anywhere near such precise measurements but we have the facility to make very accurate measurements of time should we so desire. If the accuracy we need is of the order of a half second then a stopwatch is normally sufficiently accurate. If for example we wish to know the time taken to boil a kettle full of water (about 3 minutes) we ought to be able to produce a result within 1 per cent accuracy using a stopwatch. The main limitation in this method of time measurement is the personal one of introducing errors in starting and stopping the watch.

The timing by stopwatch of the running of the mile at an athletics meeting is often acceptable as an error of perhaps a fifth of a second at both the beginning and end of a race lasting 4 minutes or more is of no significance (unless there is a chance of breaking a world record when one-tenth of a second could be vital). The hand-held stopwatch is no longer sufficiently accurate even in amateur sports events where there is no possibility of shattering world records if the distance run is 100 metres. A fifth of a second error at the start and finish could produce a percentage error of 4 per cent in a 10-second interval.

The main precaution to be taken therefore in time measurement (especially when the time interval is small) is the starting and stopping of the timing mechanism.

The personal error is removed if the event being timed is itself responsible for the starting and stopping of the clock. Photoelectric cells are frequently employed for this purpose.

The time taken for a mass to fall a distance of say 3 metres can easily be measured accurately to within 1/1000 of a second or better by the apparatus shown in Fig. 4.21. The electronic oscillator is frequency controlled by the use of a quartz crystal. This gives a constancy of frequency about a known value of better than 1 part in 10^5 (often 1 in 10^6). The

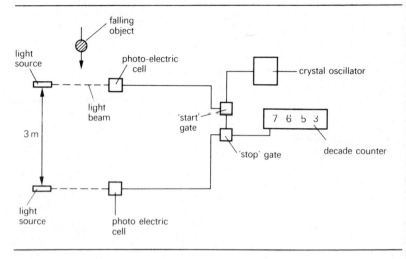

Fig. 4.21   Measurement of a short time interval.

oscillations can be easily counted electronically. The 'gate' from the oscillator is closed when the first pulse is received from the upper photo-electric cell as the mass breaks a light beam. The 'gate' is opened when the mass passes the second cell. The indication on the counter is the number of cycles which occur in the time interval between the opening and closing of the gates. This is directly proportional to the time being measured. If for example the oscillator has a frequency of 10^5 Hz then each cycle represents a time interval of 10^{-5} seconds or 0.01 ms. If the counter used employs a five-digit display then the final digit represents 0.01 ms. The display will read up to 0.99999 seconds.

There is always a ±1 digit error in this type of measurement. The opening and closing of the gates can each result in an error of 1 cycle. The oscillator itself may also have an accuracy of only 1 part in 10^5 which corresponds to 1 digit. The total cumulative error could therefore be ±4 digits or ±0.04 ms, which is still very small. For example, the display might indicate 90857 so that the interval of time would lie between 0.90861 and 0.90853 seconds.

If an accuracy with less than 1 ms error is required a four-digit display would be adequate.

The electronic counting is carried out with a series of 'flip-flop' circuits which switch from one state to another as each positive half cycle of the oscillations is applied. This circuit effectively divides the input frequency by 2. If the frequency of the oscillator is 10^5 Hz then the output from the first flip-flop is 50×10^3 Hz, the next 25×10^3 Hz and so on. This is essentially a binary division. By feeding back pulses in the frequency dividing networks it is possible to transform the binary counting to denary. The display of digits in modern frequency counters is often by means of light-emitting diodes (which are used extensively in electronic hand calculators).

Summary

Measurements of heat quantities employ thermometers and calorimeters. The calorific value of solid fuel is measured by means of a 'bomb' calorimeter.

Light measurements are based on the candela but are to a certain extent subjective.

Sound measurements tend to be subjective, and although a quantitative basis of sound is the power transmitted per unit area, the phon is employed as the subjective measurement. Ratios of sound power employ a logarithmic scale and use the decibel.

Fluid measurements are usually centred on pressure (where manometers may be used) or flow (where either rotors or venturi tubes are used). Air flow can be measured by pitot tubes.

Time can be measured very accurately using electronic counters.

Questions

To what extent is heat a subjective measurement?

How does humidity affect the temperature of the body?

What happens when the body temperature falls?

What is meant by the term 'colour blindness' and how is it detected? Do animals have the same visual ability as humans?

To what extent can the eye detect angular differences?

What units does an optician use?

What level of sound is experienced in many discotheques?

Draw up a table of sound levels (in phons) from the proverbial 'pin dropping' to the noise of a jet aircraft, inserting many commonly heard sounds, e.g. pneumatic drills, traffic noise.

Orifice plates are sometimes used in flow measurements. How do they operate?

What is a water meter and where is it used?

How does the 'gas meter' work?

What are the calorific values of petrol, paraffin, coal and wood?

How does the 'radar speed trap' used by the police function?

What is the 'Doppler effect'?

Examples

1. What is the Fahrenheit scale of temperature?
   What is the Fahrenheit equivalent of $0°$ C $100°$ C, $200°$ C?

At what temperature does the Fahrenheit thermometer read the same as the Celsius thermometer?
Ans. $32°$ F, $212°$ F, $392°$ F; $-40°$ C or $-40°$ F

2. Define the kilo calorie.
A block of aluminium of mass 0.2 kg has its temperature raised to $100°$ C. It is transferred without heat loss to an oil bath whose temperature is $15°$ C. The mass of oil is 0.5 kg. If the specific heat of the oil is 6000 J/kg/$°$ C and that of aluminium is 8960 J/kg/$°$ C, what is the final temperature of the oil if all other heat losses are ignored?
Ans. $41°$ C

3. A lamp emits 500 lumens and its luminous intensity is uniform. What is the illumination of a plane surface of area 1 cm^2 held normal to the light rays at 1.5 m from the source?
Ans. 0.00176 lux

4. The sound from a loudspeaker is measured on a meter calibrated in decibels. A deflection indicating 50 dB is noted. Three more speakers with the same sound output are now placed alongside the first. What is the new dB reading?
Ans. 56 dB

5. What is the sound intensity level in decibels at a sound intensity of 10^{-5} W/m^2? What is the measurement in phons at this intensity at (*a*) 100 Hz, (*b*) 10 kHz?
Ans. 70dB;   60 phons, 58 phons

6. The sound level at a certain point falls from 78 dB to 43 dB in 0.5 seconds. What is the reverberation time?
Ans. 0.86 s

7. The pressure in a vessel is measured using an open U tube manometer containing mercury. The level of the mercury in the side of the tube nearer the vessel is 32.5 cm and in the other side is 16.2 cm. What is the pressure in the vessel? (Density of mercury = 13.6 x 10^3 kg/m^2.)
Ans. 79,300 N/m^2

Chapter 5

Electronics I (analogue circuits)

5.1 Introduction

Electronics is a fast-changing branch of technology. A mere two decades ago the thermionic valve dominated the scene although the transistor had made an appearance and was rapidly gaining ground. Today the transistor as a discrete element is, in its turn, being replaced by the integrated circuit. Nevertheless, the thermionic valve still survives and has certain advantages over the solid state devices in the regions of extremely high frequencies and large powers. It is still extensively employed in the form of the cathode ray tube. Since the action of a simple thermionic valve is possibly easier to follow than that of a semiconductor device it will be dealt with first. Both devices employ moving charges in their operation but in the latter case the charges remain always within the semiconductor material.

The chapter title also deserves some explanation. Electronic circuitry tends to fall into two categories: (i) where voltage or current waveforms are being considered whose amplitudes vary with time in sometimes very complicated ways and (ii) where voltage or current waveforms are being considered which are square shaped. In the former case the circuit is usually designed to deal with a waveform in such a way that a faithful reproduction of it is possible. This is typical of a 'hi-fi' system.

In the latter case the circuit is usually concerned with the determination of whether a voltage (or current) is present or not.

The latter circuits are used in logic situations and will be dealt with in Chapter 6. The genus of these circuits is referred to as Digital Circuits. The former circuits (dealt with in this chapter) are called Analogue Circuits. Circuits in this field are usually analysed by considering mainly sinusoidal voltages (or currents). In analogue circuits, waveform is of considerable importance and we tend to consider sine waves of varying frequencies. One is said to be operating in the frequency domain

With digital circuits we are usually interested in the instant in time when square pulses appear. This is called the area of time domain.

5.2 The thermionic diode

The simplest of thermionic devices employs two electrodes and is termed the diode. It consists of a heated cathode surface, surrounded by a metal

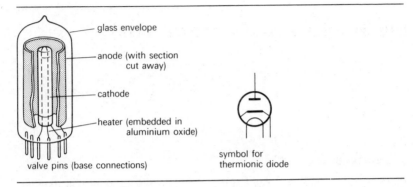

glass envelope

anode (with section cut away)

cathode

heater (embedded in aluminium oxide)

valve pins (base connections)

symbol for thermionic diode

Fig. 5.1 The thermionic diode.

cylinder (the anode), the assembly being mounted in a glass envelope from which all gas has been excluded (see Fig. 5.1).

The cathode is normally made of nickel which is thinly coated with barium and strontium oxides. The heater for the cathode is of tungsten and is electrically insulated from the cathode. When the heater is connected to a suitable low voltage supply (4—6 V being usual) the heat generated raises the cathode surface to about $1000°C$ (1300 K). Under these conditions the cathode surface begins to lose electrons as the electrons present in the cathode are thermally agitated sufficiently to be 'thrown out' from the cathode.

Since the cathode before being heated was electrically neutral, i.e. contained an equal number of positive and negative charges, the emitted electrons being negatively charged leave behind a cathode with an excess of positive charge. This charge tends to attract back to the cathode the emitted electrons and a cloud of electrons develops around the cathode, some leaving and some returning to the surface. Since the electrons leave the cathode with random velocities centred around some mean value the density of the electron cloud varies taking on a Gaussian distribution. This cloud is termed a space charge and reaches its maximum density a millimetre or so from the cathode surface.

Under these conditions very few electrons reach the cylindrical anode.

If however, the anode is made positive to the cathode by applying a potential difference between the cathode and anode a large number of the electrons in the space charge reach the anode and give up their charge. The effect is a flow of electrons from cathode to anode and a corresponding flow of conventional current in the opposite direction (Fig. 5.2). Increasing the anode voltage in a positive sense causes a modification of the space charge and increases the current flow. A point of saturation is reached when the anode voltage has virtually removed the space charge and all the emitted electrons from the cathode then reach the anode. Further increase in the voltage causes little change in current and the only way that further increase in anode current can take place is to increase the

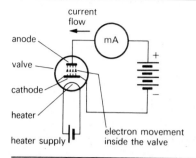

Fig. 5.2   Current flow through a diode.

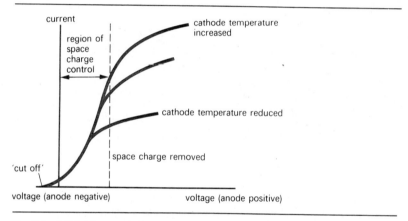

Fig. 5.3   Diode characteristics.

number of electrons emitted by the cathode. This is achieved by raising its temperature.

If the anode is made negative with respect to the anode electrons are repelled from the anode and no current flows.

The characteristics of the diode are shown in Fig. 5.3. In the conducting direction the valve behaves as a non-linear resistance. With the anode negative, current ceases. The diode therefore behaves as a rectifier, and if connected in series with an a.c. supply and a resistor will only allow current to flow during the positive half cycles (Fig. 5.4). The result is half-wave rectification.

Full-wave rectification is achieved using two diodes and a transformer with two equal secondary windings. Referring to Fig. 5.5 when the anode of diode *a* is positive, current flows in the direction shown. In the next half cycle the anode of *a* is negative and current does not flow. Diode *b*, however now has a positive anode and passes current in the negative half cycle. Normally the diodes are operated in the region of space charge control where their effective resistance is low.

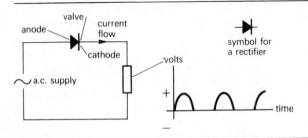

Fig. 5.4   Half-wave rectifier.

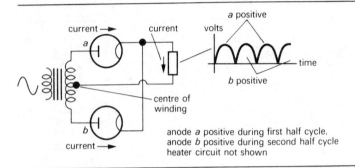

Fig. 5.5   Full-wave rectifier.

Thermionic diodes are employed where relatively high voltages are encountered — around a few hundred volts.

5.3 Semiconductors

Solid state electronic theory is based upon the property of semiconduction. Semiconducting elements germanium and silicon feature in the vast

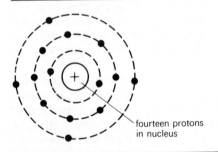

Fig. 5.6   The silicon atom.

majority of solid state devices. As the name implies, a semiconductor has a conductivity resting between that of an insulator and that of a conductor, typically 0.01 Ω-m. In order to understand how conduction takes place in a semiconductor it is necessary to know how the atoms forming a germanium or silicon crystal are arranged. Both these elements have a valency of 4; that is, the electrons in the outermost ring of electrons surrounding the atomic nucleus number 4. Taking silicon as an example, silicon has the atomic number 14, in other words each atom possesses 14 electrons. These are arranged in rings of 2, 8 and 4 (see Fig. 5.6). The outermost electrons in each atom link with corresponding electrons of adjacent atoms to form a tetrahedron crystal. The linking of the electrons is referred to as covalent bonding (Fig. 5.7). It is rather easier to draw a crystal lattice of this sort as a flattened tetrahedron as in Fig. 5.8.

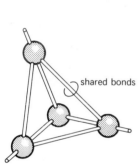

shared bonds

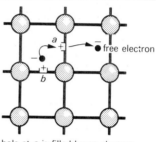

free electron

hole at *a* is filled by an electron liberated by the bond breakage at *b*. The hole at *a* has apparently moved to *b*.

Fig. 5.7 Crystalline structure of silicon showing covalent bonds.

Fig. 5.8 Hole and electron movement in silicon or germanium.

Under very low temperature conditions the bonds remain in position but as the temperature is increased some of these bonds break, each releasing an electron which remains within the crystal but is free to move. Mobile electrons are necessary for conduction and exist in vast numbers in a conductor. In a semiconductor, however there are relatively few at room temperature but the application of a little heat causes more bonds to break, releasing more electrons and raising the conductivity of the material.

When an electron is released it leaves behind an atom with a positive charge. These regions of positive charge are referred to as holes and are fixed in position. It is, however, possible for a mobile electron to be attracted to a hole re-establishing the bond. If a hole appears in one atom, a free electron from a neighbouring atom may fill this hole and to all intents the hole has moved. We can therefore consider holes to be mobile positive charges although due to their very nature they are not as mobile as free electrons and in fact only give the impression of movement (see Fig. 5.8).

Pure silicon (or germanium) therefore produces both mobile negative charges (electrons) and apparently mobile positive charges (holes) and these will occur in pairs. Increasing the temperature will generate additional hole-electron pairs. It is convenient to consider holes produced in this way and electrons as separate 'particles' of opposite polarity which both contribute to the conduction of electricity.

5.4 Conduction in semiconductors

If a p.d. is placed across a semiconductor at room temperature positive holes will move to the negative end and negative electrons to the positive end. The effect is cumulative and the result is a flow of conventional current (see Fig. 5.9). This action is known as intrinsic conduction. The

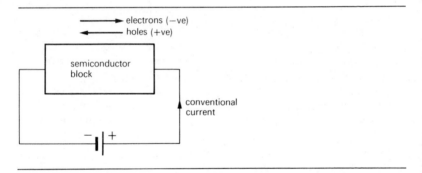

Fig. 5.9 Conduction in an intrinsic semiconductor.

semiconductor must not only be very pure for this effect to be observed but the crystal lattice must be as perfect as possible. It is necessary to manufacture semiconductors very carefully indeed to achieve this result.

If a very small quantity of some impurity element having a valancy of 5 (i.e. 5 electrons in the outermost ring) is introduced into the lattice of the semiconductor crystal with valency 4 the new atom tries to fit in as best it can but this leaves a spare electron which is very loosely attached (see Fig. 5.10). In consequence the free electron readily leaves the parent atom, leaving behind a hole which will not readily capture an electron. Unlike the pure semiconductor, the hole is not mobile, so that we now have an impure semiconductor which produces mobile electrons without the corresponding mobile holes. Such a semiconductor is said to be doped (arsenic added to germanium produces this result). The production of the mobile electrons in such a case is not very dependent upon temperature but it must be remembered that the production of hole-electron pairs by intrinsic action occurs simultaneously. Normally the extrinsic action, as it

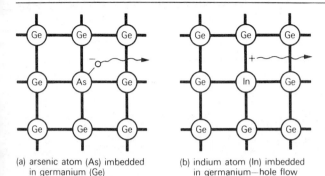

(a) arsenic atom (As) imbedded
   in germanium (Ge)
   —electron flow

(b) indium atom (In) imbedded
   in germanium—hole flow

Fig. 5.10   Silicon behaves in the same fashion.
(*a*) n-type semiconductor.
(*b*) p-type semiconductor.

is called, of producing mobile electrons only completely swamps the intrinsic action but obviously as the temperature is raised the latter becomes more important. At room temperature the extrinsic action dominates. Semiconductors which have been doped in this way are called 'n type' to signify production of negative mobile charges.

If on the other hand a trivalent element such as indium is added to pure germanium the foreign body tries to fit in to the crystal lattice but leaves an incomplete bond resulting in a positive hole. Like the n type material this hole is very loosely attached and is mobile leaving behind a fixed seat of negative charge (Fig. 5.10).

Such material is referred to as 'p type' to signify production of positive mobile charges. Summarising, n type silicon or germanium produces mobile electrons and fixed positive charges. p type material produces mobile holes and fixed negative charges. In both cases generation of hole electron pairs also occurs but on a much smaller scale.

5.5 The pn junction

If a piece of p type silicon (or germanium) is placed in extremely close contact with n type silicon (or germanium), the following action occurs.

The mobile holes in the p type semiconductor diffuse across the junction into the n region and the mobile electrons in the n type semiconductor diffuse across the junction into the p region. This creates a potential barrier since the n region now has an excess of holes and the p region now has an excess of electrons. The n region is positively charged and the p region is negatively charged. This potential difference (typically 0.7 V in germanium) is sufficient to prevent any further diffusion of mobile charges.

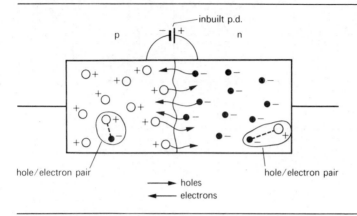

Fig. 5.11   Mobile charges in a pn junction.

If an external p.d. is now applied in the same sense as the existing potential barrier the only possible movement of charge is that due to the minority charges (the electron pairs generated by thermal means). Under these conditions the pn junction is said to be reverse biased and further small increases of voltage causes no further increase in current (Fig. 5.12). If however the voltage is considerably increased a breakage of bonds occurs in the crystal lattice producing an increase in mobile charges and an avalanche effect is generated, causing a rapid increase in current.

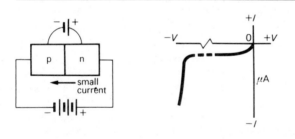

Fig. 5.12   Reverse bias conditions.

If the polarity of the applied voltage is reversed then the potential barrier is removed and the majority charges (extrinsic action) can now easily cross the pn junction. There is a rapid increase in current as the voltage is increased and the junction is said to be forward biased (Fig. 5.13).

The total characteristic of the device is shown in Fig. 5.14 and it is evident once more that we have a rectifier.

The pn junction has similar properties to the thermionic diode but no

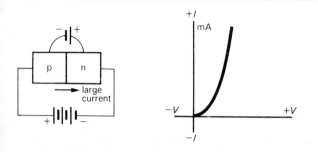

Fig. 5.13 Forward bias conditions.

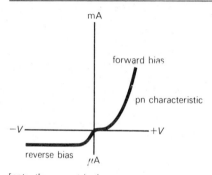

[note: the current in the reverse bias direction is plotted on a magnified scale (× 1000).]

Fig. 5.14 pn junction characteristic.

cathode heating supply is necessary. Such junctions can be made very cheaply and occupy very little space. If they are to be used where large currents are anticipated it is necessary to provide some means of cooling them. A piece of metal is attached to the device offering a large cooling area. Such arrangements are called heat sinks.

5.6 The transistor

Basically the transistor consists of two pn junctions in series forming a pnp or an npn sandwich (Fig. 5.15). The centre section is made as thin as possible. Before the application of any external potential differences two potential hills are established as with the pn junction just described.

If the pnp device is considered the transistor so formed normally has the first junction forward biased and the second junction reverse biased.

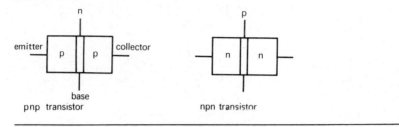

Fig. 5.15   Diagrammatic representation of the transistor.

The left-hand p region is called the emitter and the right-hand p region the collector. The n type section is the base.

Holes in the p type emitter cross the first junction fairly easily since it is forward biased. The second barrier, although reverse biased for holes in the right-hand region, is forward biased for holes in the central section. Consequently, nearly all the holes which cross the first pn junction cross the second np junction. A few may recombine with electrons in the n section but typically some 99 per cent of the holes from the emitter cross the central region and appear at the collector (Fig. 5.16). The forward bias

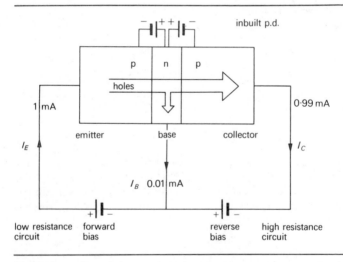

Fig. 5.16   Action of the transistor (common base). Movement of majority carriers (holes in pnp transistor).

action of the first junction provides a relatively low resistance circuit but the reverse bias across the second junction produces a high resistance circuit. The main change which has been produced by the emitter/base voltage is that holes generated from a low resistance circuit appear in the collector which is a high resistance circuit.

If a resistance of a few thousand ohms is placed in series with the collector the current crossing the base/emitter junction will produce a p.d. across this resistance. Let us illustrate by an example.

Let the emitter/base be suitably biased and assume an alternating p.d. of 50 mV be applied in series with the bias voltage. This alternating voltage may produce an alternating current of 0.1 mA in the emitter. If 99 per cent of the current in the emitter reaches the collector the collector current is then 0.099 mA. Assuming that a resistance of 10 kΩ is put in the collector circuit the 0.099 mA produces a voltage of 990 mV across it. The output voltage is 990 mV compared with an input voltage of 50 mV, i.e. a voltage gain of 19.8 has occurred.

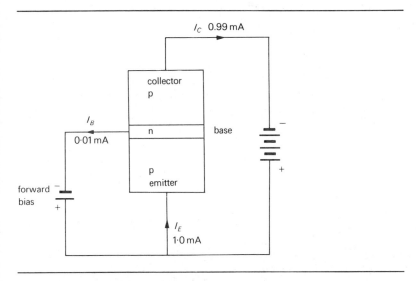

Fig. 5.17   Common emitter connection.

The ratio of collector current to emitter current when no added resistance is present is given the symbol α. In this example α is 0.99. Although this method of operation can be used for amplification it is more usual to interchange the connections between base and emitter so that the emitter is common to the input and output circuits (Fig. 5.17). This has one significant advantage, the input circuit now has a higher resistance since the base current is much lower than the emitter current (1 per cent of I_E in the example quoted). Thus the input resistance of the common emitter amplifier is about 100 times as great as the input resistance of the common base amplifier.

The current ratio in the common base configuration; $\dfrac{I_C}{I_E} = \alpha$

In the common emitter configuration the ratio needed is $\dfrac{I_C}{I_B} = \beta$

But $I_B = I_E - I_C$ and $I_E = \dfrac{I_C}{\alpha}$

$\therefore \quad \beta = \dfrac{I_C}{\dfrac{I_C}{\alpha} - I_C} = \dfrac{\alpha}{1 - \alpha}$

Hence if $\alpha = 0.99$ $\beta = 99$

i.e. in common emitter circuits there is a current gain.

5.7 The transistor amplifier

Transistors can use pnp sandwiches or npn sandwiches. In the latter case conduction takes place mainly by electrons rather than holes and the polarity of the applied voltages has to be reversed (see Fig. 5.18). The symbols for the two transistors are given.

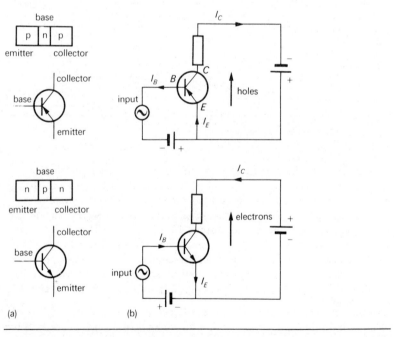

Fig. 5.18(*a*)   pnp and npn symbols.
 (*b*)   comparison of pnp amplifier and npn amplifier.

The most useful characteristics of the transistor are the output (or collector) characteristics. These are obtained by varying the voltage across the collector/emitter (V_{CE}) and observing the collector current (I_C) while at the same time maintaining the base current (I_B) constant. The starting point is with $I_B = 0$ and the curve obtained is a very flat one typical of a reversed biased pn junction. As I_B is increased the characteristics assume a slightly steeper slope. (Generally speaking a capital letter as the suffix of a quantity, e.g. V_{CE} indicates a d.c. or steady value. Lower case letters refer to a.c. or changing quantities, e.g. I_e or i_e implies a changing emitter current.)

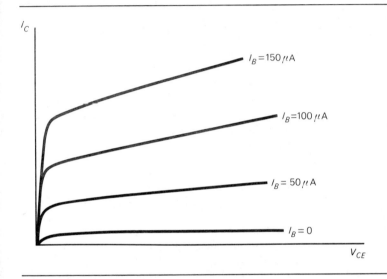

Fig. 5.19 Collector characteristics of a transistor in common emitter (CE) connection.

Figure 5.19 shows a sample for a transistor connected as a common emitter circuit. The transistor can be used as an amplifier by connecting a resistor in the collector circuit. When a resistor (a load resistor) is connected in series with the collector and the supply voltage, the collector/emitter voltage depends upon this supply voltage, the collector current flowing and the value of the load resistance

$$V_{CE} = V_{BB} - I_C R_L \qquad V_{BB} = \text{supply voltage (usually about 10 V)}$$
$$I_C \quad = \text{collector current}$$
$$R_L \quad = \text{load resistance}$$

This is a graph of a straight line of negative slope R_L and passing through the point $V_{CE} = V_{BB}$. Such a line is called a load line and is shown in Fig. 5.20. If the base/emitter bias voltage is adjusted to provide a suitable base current (point Q on the diagram), the application of an alternating voltage

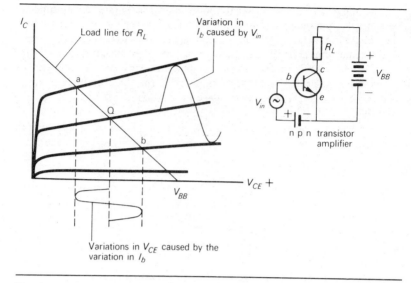

Fig. 5.20   Action of the CE npn transistor amplifier.

between base and emitter can cause the base current to swing between limiting values of base current. This in turn causes the collector/emitter voltage to move up and down the load line (a–b). The output voltage is then found by projecting vertically downwards on the V_{CE} axis.

The complete amplifier circuit for a npn transistor is shown in Fig. 5.21. R_1 and R_2 are adjusted to provide the necessary bias voltage for the base establishing the quiescent operating point (Q). C_1 is an input capacitor allowing the alternating input voltage to be applied without disturbing the bias conditions. Similarly C_2 is an output capacitor which

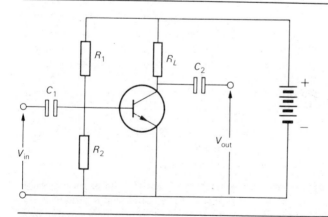

Fig. 5.21   Practical CE npn circuit.

allows the alternating component of the collector voltage to be obtained while 'blocking off' the d.c. component.

The reactance of C_1 and C_2 should be small at the operating frequency. Two further points must be made:

1. The output voltage is out of phase (180°) with the input.
2. Since the spacing of the output characteristics are never uniform a sinusoidal variation in I_b cannot produce a pure sine wave of collector voltage. In other words, some distortion is present.

For many applications of transistor amplifiers it is not necessary to use the graphical approach and a 'black box' approach is often much quicker and sufficiently accurate for most applications. This approach has the additional advantage in the fact that it is perfectly general and can be applied to common base or common emitter circuits. The transistor is treated as a device which has input terminals a, b and output terminals c, d. It is assumed that alternating voltages v_i and v_0 can be applied to the input and output terminals and corresponding currents i_i and i_0 flow (Fig. 5.22). If we take the special case when terminals c and d are connected then the application of a voltage v_i to the input causes a current i_i to flow and $v_0 = 0$.

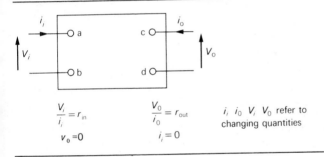

Fig. 5.22   'Black box' approach.

The ratio $\dfrac{v_i}{i_i}$ = input resistance and is given the symbol r_{in} or h_i

h_i is a resistance

Similarly if we apply a voltage to the output and note the current flowing in the output circuit when the input is not connected to anything we obtain the output resistance

$$\frac{v_0}{i_0} = r_{out}$$ v_1, v_0, i_i and i_0 all refer to alternating or changing quantities.

This ratio can be derived from the collector characteristic. The slope of the characteristics is

$$\frac{\delta i_c}{\delta v_c} = \frac{1}{r_{\text{out}}} \quad \text{or } h_o$$

h_o is a conductance.

Finally we can obtain the ratio of input current and output current with the output on short circuit, i.e. when $v_0 = 0$. This is the current gain of the transistor — either α or β.

We then give this the symbol h_f.

h_f is a ratio without any associated units.

These three parameters, h_i, h_o and h_f, are called hybrid parameters — so called because one is a resistance, one a conductance and one a ratio. If we are dealing with a common base circuit the parameters become h_{ib}, h_{ob}, h_{fb}.

For a common emitter circuit they become h_{ie}, h_{oe}, h_{fe}. The suffix indicates the method of connecting the transistor. (A fourth parameter, h_{re}, a voltage ratio, also exists but is usually of negligible proportions. It will be disregarded at this stage.)

The equivalent circuit of a common emitter circuit thus becomes that shown in Fig. 5.23. A current generator βi_b or $h_{fe} \cdot i_b$ is in the output

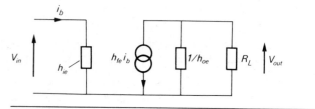

Fig. 5.23   Transistor common emitter equivalent circuit.

circuit with a resistance $1/h_{oe}$ in shunt. The output voltage is approximately $-\beta i_b R_L$, if R_L is not too large and h_{oe} is fairly small. The output voltage is shown with a negative sign to indicate the phase reversal.

Let us consider a numerical example.

Example 5.1

What is the output voltage for a transistor with a load resistance of 5 kΩ and an input of 50 mV if the h parameters are
$h_{ie} = 20 \text{ k}\Omega \quad h_{oe} = 50 \text{ μS} \quad h_{fe} = 99?$

$$i_b = \frac{v_i}{h_{ie}} = \frac{50}{20} \text{μA} = 2.5 \text{ μA} \quad \text{(See Fig. 5.23.)}$$

the current in the output circuit = $h_{fe} \times i_b = 99 \times 2.5 = 248 \text{ μA}$

Current divides in the output in ratio $\dfrac{1}{h_{oe}}$ and R_L

i.e. in ratio 20 kΩ to 5 kΩ or 4 : 1.

$\therefore \dfrac{4}{5}$ of the current of 248 μA flows through R_L

$\dfrac{4}{5} \times 248 = 198 \ \mu$A

$\therefore V_{out} = 198 \ \mu$A $\times R_L = 198 \times 5 = 990$ mV

In this case the effect of h_{oe} was not negligible.

The parameters h_{oe} and h_{fe} can be obtained directly from the collector characteristics (see Fig. 5.24). h_{ie} must be derived from the input characteristics (i_b/v_{be}). It will be appreciated that the hybrid parameters are not constant for a given transistor but depend upon the point on the characteristic at which they are measured. For typical operating conditions where only small amplitude voltages and currents are considered the variations are not usually very significant.

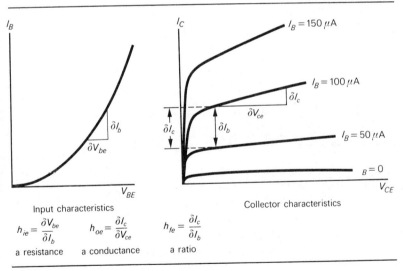

Input characteristics   Collector characteristics

$h_{ie} = \dfrac{\partial V_{be}}{\partial I_b}$ $h_{oe} = \dfrac{\partial I_c}{\partial V_{ce}}$ $h_{fe} = \dfrac{\partial I_c}{\partial I_b}$

a resistance   a conductance   a ratio

Fig. 5.24   'h' parameters derived from the characteristics.

When the amplifier is connected to other amplifiers or circuits the various direct voltages form the bias or main supply must normally be blocked off so that only the a.c. signal is passed on. Although the d.c. components are vital for the operation of the transistor they do not constitute part of the wanted signal. The input and output capacitors are provided for this purpose (Fig. 5.25). This obviously modifies the action of the amplifier since capacitors have a reactance $1/\omega C$ which varies with frequency.

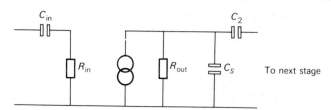

R_{in} = effective input resistance including bias resistors

R_{out} = effective output resistance R_L in parallel with $1/h_{oe}$

Fig. 5.25   The fuller equivalent circuit including capacitors.

This causes the gain of an amplifier to fall at low frequencies since the input now consists of an R/C circuit. The input voltage to the base is

$$\frac{v_{in}R_{in}}{\sqrt{R_{in}^2 + \left(\dfrac{1}{\omega C_{in}}\right)^2}}$$

where R_{in} is h_{ie} effectively in parallel with any biasing resistors

if ω is large $R_{in}^2 \gg \left(\dfrac{1}{\omega C_{in}}\right)^2$

so that $v_{be} = v_{in}$ very nearly.

At low frequencies $\dfrac{1}{\omega C_{in}}$ is no longer negligible and $v_{be} < v_{in}$.

At the particular value $\dfrac{1}{\omega C_{in}} = R_{in}$ the base voltage is $\dfrac{1}{\sqrt{2}} v_{in}$ or $0.707\, v_{in}$.

The frequency at which this occurs is arbitrarily taken to be the lower frequency limit of the operation of the amplifier.

A similar situation occurs at a much higher frequency where stray capacity effects shunt the output. If the stray shunt capacitance is C_s then the output voltage falls to 0.707 of the value of the mid-frequency when $\dfrac{1}{\omega C_s} = R_{out}$. R_{out} is composed of $\dfrac{1}{h_{oe}}$ and R_L in parallel. Figure 5.23 may help to illustrate this.

A typical response curve for a transistor amplifier with a collector resistance load is shown in Fig. 5.26. The two frequencies define the bandwidth of the amplifier.

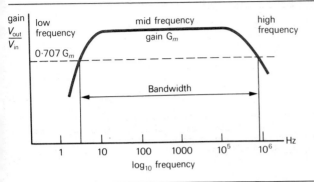

Fig. 5.26   Frequency response of an amplifier.

Summary

Electronic circuits divide into two main groups, analogue and digital. Analogue circuits deal essentially with sine waves of varying frequency.

Although thermionic devices are still in fairly widespread use the transistor has replaced the thermionic valve in many applications.

The pn junction behaves similarly to the thermionic diode, both are essentially rectifiers. The pnp or npn device is an active circuit capable of amplification. The collector characteristics may be used to determine the output for a given input. Alternatively the hybrid parameters allow an algebraic approach. Amplifiers in general have a bandwidth of operation.

Questions

Nowadays the junction transistor is being replaced by the F.E.T. in many instances. What is an F.E.T.?

Integrated circuits are beginning to be used extensively. What is an integrated circuit?

How is a pn junction made?

Can semiconductor devices operate in environments which are (i) hot (ii) radioactive?

What precautions should be observed when soldering semiconductor devices into circuits?

Which are more reliable — valves or transistors? Substantiate your answer.

Silicon has largely replaced germanium as a semiconductor. Why?

How is silicon purified to the high degree of purification required by semiconductors?

Can resistors or capacitors be made by semiconductor techniques?

Examples

1. A diode has a constant forward resistance of 60 Ω and an infinite reverse resistance. It is placed in series with a 100 Ω resistance and a 50 Hz sinusoidal supply of peak value 50 V.

What is (*a*) the average current flowing?
   (*b*) the average voltage across the resistor?
   (*c*) the average voltage across the diode?
Ans. 100 mA;   10 V;   −10 V

2. A thermionic diode has the following characteristics:

V_a (V)	0	20	40	60	80	110	130	150	180
I_a (mA)	0	2.3	5.6	9.2	12.9	18.5	21.6	23.7	25.0

It is placed in series with a 1 kΩ resistor and a d.c. supply (anode positive). What current flows and what is the p.d. across the circuit if the p.d. across the diode is 100 V? If the heater supply were switched off, how would the values of the current and the p.d. across the diode be affected?
Ans. 17.2 mA   117 V;   0 mA   117 V

3. A diode may be considered to have an infinite reverse resistance but a straight line forward characteristic passing through the point $v = 0$, $i = 0$ and $v = 25$ V, $i = 50$ mA.
   It is connected in series with a resistance of 1.5 kΩ and a 50 V, 50 Hz a.c. supply.
   A moving-coil voltmeter whose resistance is 1.5 kΩ is connected (*a*) across the resistance, (*b*) across the diode. Sketch the waveforms of voltage and calculate the indications on the meter.
Ans. 13.5 V   −6.75 V

4. A battery is to be charged from a 23 V, 50 Hz supply via a charging regulator and a rectifier having ideal characteristics. Draw the circuit indicating clearly the direction of current and the polarity of the battery.
   What is the time in seconds over which the rectifier conducts in each cycle?
   The e.m.f. of the battery is 24 V.
Ans. 4.7 ms

5. Draw the waveform of the output voltage obtained from the circuit shown. Each diode when conducting can be regarded as having a constant resistance of 1 kΩ V_{in} is a ramp voltage increasing at a constant rate of 1 V/s.

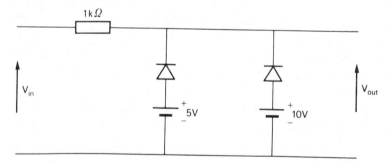

6. The collector characteristics of a pnp transistor in common emitter connection may be considered as three straight lines through the following points:

(*a*) $I_C = -7.3$ mA $V_{CE} = -1.0$ V; $I_C = -8$ mA $V_{CE} = -9$ V   ($i_b = 80$ μA)
(*b*) $I_C = -3.4$ mA $V_{CE} = -1.0$ V; $I_C = -4$ mA $V_{CE} = -9$ V   ($i_b = 40$ μA)
(*c*) $I_C = -0.2$ mA $V_{CE} = -1.0$ V; $I_C = -0.4$ mA $V_{CE} = -9$ V   ($i_b = 0$ μA)

What are the values of α, β and the effective collector circuit resistance?
   If the supply voltage is −10 V and the load is 1.25 kΩ find the current gain of the circuit.
Ans. 0.99   100;   13 kΩ   85 times

7. The h parameters of a transistor are h_{ie}, $= 10 \text{ k}\Omega$, $h_{fe} = 99$, $h_{oe} = 80 \text{ }\mu\text{S}$. Find the voltage gain of a common emitter amplifier which employs this transistor and a collector load resistance of $7.5 \text{ k}\Omega$.

If two such amplifiers are connected in cascade (i.e. one after the other), find the total voltage gain.

Draw the circuit for the two-stage amplifier including the bias resistors and the capacitors and indicate practical values if the circuit is to operate at a fixed frequency of 0.5 kHz.

Ans. 45.8   1450.

8. The output characteristic of a transistor connected in the common emitter configuration is given below. Plot the characteristic and use it to determine the collector current and the collector-emitter voltage when the transistor is operating with a load resistor of $3 \text{ k}\Omega$, a voltage supply of 10 V and is biased with a base current of $40 \text{ }\mu\text{A}$.

If an alternating signal of peak to peak value 0.2 V is applied to the base such that the base current alternates between $20 \text{ }\mu\text{A}$ and $60 \text{ }\mu\text{A}$, find the output current swing, the output voltage swing and the current gain.

Collector voltage (volts)	Collector current (mA)			
	For base current = $20 \text{ }\mu\text{A}$	For base current = $40 \text{ }\mu\text{A}$	For base current = $60 \text{ }\mu\text{A}$	For base current = $80 \text{ }\mu\text{A}$
3	0.91	1.6	2.3	3.0
5	0.93	1.7	2.5	3.25
7	0.97	1.85	2.7	3.55
9	1.00	2.05	3.0	4.05

Find also the h parameters and recalculate the output voltage swing from the equivalent circuit. Why do the results obtained from the graph not agree completely with the calculated results?

Ans. 1.7 mA   4.9 V;   1.3 mA   3.9 V   32;
$h_{ie} = 5 \text{ k}\Omega$ $h_{oe} = 75 \text{ }\mu\text{S}$ approx. $h_{fe} = 45$ approx.
(*Crawley College*)

Chapter 6

Electronics II (logic circuits)

6.1 Problems and their reduction to logic statements

Being presented with a problem is a situation with which we are all familiar. The problem may be a mathematical one which is capable of exact solution and may have only one correct answer, e.g. solve $2x + 17 = 4$. The solution is then said to be unique. Mathematical problems sometimes generate more than one solution, e.g. solve $y^2 + 2y - 7 = 0$.

Everyday problems are seldom of this type. There is frequently no completely correct solution since what is right or wrong may be a matter of opinion. Even what is considered by the majority as the correct solution today may be judged incorrect in a month's time in the light of additional information.

Often the problem is very complex and there is no easy route to any solution. For example, the problem generated by an economic crisis is one which appears to defy solution by successive governments.

Nevertheless any problem, however complicated, can be broken down into smaller problems or steps and often the alternative solutions to these more elemental problems are limited. Complex problems can sometimes be solved in this way. It is possible in many instances to so frame questions within a problem that only two alternatives are possible — an answer 'Yes' or an answer 'No' and the 'Don't know' answer is literally 'out of the question'.

Take a very trivial problem. 'What am I going to do tonight as I appear to have some free time?' Perhaps an unlikely question to many students but one which demands an answer.

The answer might be 'I think I will visit the "local"', but that in turn may depend upon the weather and financial resources (although one could still visit the pub in the fond hope that some kind friend will buy one a drink). The problem can now be split down so that the alternatives are more clearly seen, e.g. 'I will go to the pub if it doesn't rain and I will buy a drink if my pay packet arrives in time'. Progress is now being made, the problem is being identified and the individual is being forced to take a decision.

The questions now are:

1. Is it raining?   Yes or No.
2. Has my pay packet arrived?   Yes or No.

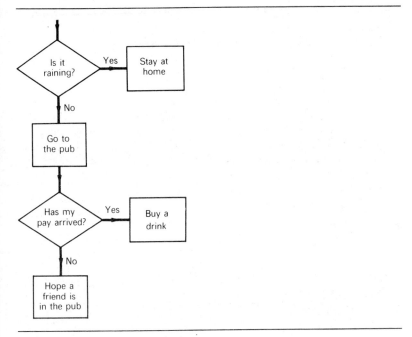

Fig. 6.1   Algorithm to decide what to do.

The decision to go to the pub rests on (1) and the decision to buy a drink rests on (2). There is now a sequence of decision-taking. This can be shown graphically by means of an algorithm (Fig. 6.1). The clear cut answers Yes or No suggest a binary device, corresponding to two clearly defined states ON or OFF.

In modern technology this is a situation which occurs time and time again and use is made of logic circuits to enable problems of this nature to be presented and solved. They are used extensively in computer circuits.

6.2 The four basic logic statements

A logic circuit is essentially a two-position device having states corresponding to ON and OFF and behaves as a switch. The two states or positions of the logic circuit can therefore correspond to the answers Yes and No. It is a binary device and the alternatives can be considered to be a one (1) or a zero (0) i.e. 1 corresponds to the ON position (Yes) and 0 to the OFF position (No).

Before we can apply logic circuits to a problem it is necessary to provide a logic statement or question which has two alternatives.

Fortunately with logic of this sort there are a very limited number of situations and possibly because of this most people find problems on logic relatively easy. (It should be pointed out that framing the problem into the right format is more than half-way to providing a solution.)

Let us consider the alternative logic statements.

(a) The OR statement

'I can obtain some additional cash if I sell either my camera or my cycle.' This is a logic statement in which three possibilities exist — if I sell **either** my camera **or** my cycle or **both** I can obtain more cash. This in logic terms is known as an OR statement.

We can represent the obtaining of cash as F. The selling of the camera as A and the selling of the cycle as B.

Thus F occurs when either A or B or both occur, i.e.

$F = A$ OR B

In logic symbols

$F = A + B$

The + sign is not an arithmetical symbol but means OR. We can represent the situation electrically by a simple circuit — two switches in parallel. The lamp lights when either or both of the switches are closed. F is the lamp when alight. A and B are the two switches (Fig. 6.2).

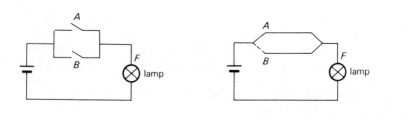

Fig. 6.2   An OR circuit.   **Fig. 6.3**   An exclusive OR circuit.

(b) The exclusive OR statement

The OR statement considered previously allowed three possibilities — one or the other or both. The exclusive OR statement allows only two. 'I am going to either the disco or the cinema but I cannot afford to go to both.'

One alternative excludes the other, there is no possibility of both. This is an exclusive OR statement and is represented by $F = A \oplus B$.

F represents the action of going somewhere, A is the action of visiting the disco, B the action of visiting the cinema.

An electrical system which is analogous to this is the two-way switch (Fig. 6.3). There is a clear alternative.

(c) The AND statement

'The River Thames flows under Waterloo Bridge and Tower Bridge.' This is an AND statement. No alternative exists since the river must flow under both bridges.

This logic statement is represented by $F = A$ and B or in symbols $F = A \cdot B$ (the dot in this case means AND).

F represents the River Thames flowing, A represents flowing under the Waterloo Bridge, B under Tower Bridge. Both events A and B must occur if the River Thames flows at all.

An electrical circuit analogous to this situation is two switches in series. Both switches must be closed before the action F (the lamp lighting) occurs (Fig. 6.4).

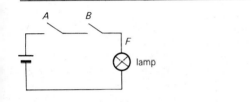

Fig. 6.4   An AND circuit.

(d) The NOT statement

This is a situation in which the exact opposite happens to that in the statement 'I am going out' is changed to 'I am **not** going out' by the addition of **not**.

The negation of an action is the NOT situation and can be expressed symbolically as follows:

$F = \overline{A}$ A means I am going out
$\overline{A}$ means I am not going out

Electrically speaking it is rather difficult to produce a simple NOT circuit. The nearest analogy is an inverter. A simple amplifier works essentially in this manner.

The algebra involving logic statements such as $F = A \cdot B$ is called Boolean algebra and should not be confused with other algebraic methods. The four basic statements and the corresponding algebra can be built up to very complex situations involving a complicated sequence of decisions.

6.3 Logic circuits – diode logic

Circuits which produce logic functions are called **gates**. These can be achieved by the use of diodes.

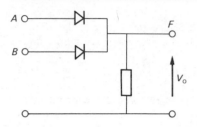

Fig. 6.5   Diode OR gate.

(a) OR circuit

A circuit which can produce the OR function is shown in Fig. 6.5. The output voltage v_0 is zero until one or both of the inputs have a positive voltage applied to it. When this happens the appropriate diode conducts and in the ideal case then acts as a short circuit. The positive input voltage now appears as v_0.

(b) AND circuit

The circuit which can produce the AND function is shown in Fig. 6.6. The input diodes are both conducting and act ideally as short circuits. v_0 is therefore small. When a sufficiently large positive voltage is applied to one or other of the inputs the appropriate diode ceases to conduct and then ideally behaves as an open circuit. When both inputs have a positive voltage of sufficient magnitude applied no further current flows through R_1 and the output voltage V_0 rises to the level of the supply (in this case 5 V).

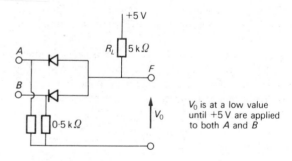

Fig. 6.6   Diode AND gate.

(c) NOT circuit

Basically the amplifier circuit behaves as a NOT circuit (Fig. 6.7). The bias voltage is such that the transistor is initially non-conducting. The output voltage V_0 is then at the positive supply potential. When a positive voltage arrives at the base the transistor then conducts and the voltage v_0 falls from its positive value.

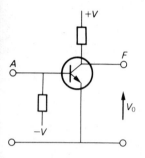

F initially at potential $+V$. When sufficiently large voltage applied to A the transistor conducts and V_0 falls to a low value

Fig. 6.7   NOT gate.

Only the simplest of electronic logic circuits have been shown here. Normally transistors are employed in conjunction with diodes forming Diode—Transistor—Logic units (or D.T.L. units) in order that the switching voltages need not be over large and to compensate for any attenuation produced by a series of logic gates. The systems described employ positive logic, i.e. a '1' corresponds to a positive voltage, a '0' corresponds to zero voltage (or a much smaller positive voltage).

6.4 Truth tables

Logic statements can be represented by Boolean algebra or by means of tables known as truth tables. The four basic logic functions, AND, OR, exclusive OR and NOT, can be illustrated in a tabular way, e.g.

AND			$F = A \cdot B$
A	B	F	
0	0	0	
1	0	0	
0	1	0	F is '1' only when A **and** B are '1's.
1	1	1	

The other statements follow in a similar manner.

	OR		$F = A + B$
A	*B*	*F*	
0	0	0	
1	0	1	*F* is '1' when either *A* **or** *B* or both are
0	1	1	set at 1
1	1	1	

	Exclusive OR		$F = A \oplus B$
A	*B*	*F*	
0	0	0	
1	0	1	*F* is '1' when either *A* **or** *B* are set at 1.
0	1	1	*F* is not 1 if *A* **and** *B* are both 1.
1	1	0	

	NOT	$F = \overline{A}$
A	*F*	
0	1	*F* is the 'opposite' of *A*
1	0	

It is of course possible to increase the number of inputs. A three-input AND gate will give an output only when all three inputs are set to 1.

The truth table is:

A	*B*	*C*	*F*	
0	0	0	0	
1	0	0	0	
0	1	0	0	
1	1	0	0	
0	0	1	0	$F = A \cdot B \cdot C$
1	0	1	0	
0	1	1	0	
1	1	1	1	

6.5 Logic symbols

Logic circuits are readily available 'off the shelf'. It is only necessary to apply the required power supply (usually around 5 V) and the circuit is active and will respond to the applied input signals. Logic circuits are usually illustrated by means of square boxes with the symbol added for an AND gate and the number 1 included for an OR gate. The NOT function is usually indicated by a square box with a small circle added (see Fig. 6.8). It is possible to join two units together such as the OR circuit and NOT circuit to produce a single NOR circuit. A circuit which will produce this is

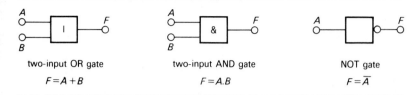

Fig. 6.8 Logic symbols.

$F = \overline{A + B}$

A	B	F
1	1	0
0	1	0
1	0	0
0	0	1

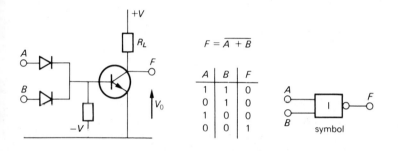

Fig. 6.9 A NOR gate using diodes and transistor (DTL).

illustrated by Fig. 6.9. The corresponding symbol is shown alongside. Initially the transistor does not conduct due to the negative voltage on the base so that no current flows through R_L and V_0 is at the supply voltage V+. Applying a positive voltage to either *A* or *B* (or both) causes the diodes to conduct, raising the potential of the base of the transistor. This in turn causes it to conduct and current flows through R_L, causing V_0 to fall to a low value (ideally zero).

A NAND gate is an AND and a NOT gate in series (Fig. 6.10).

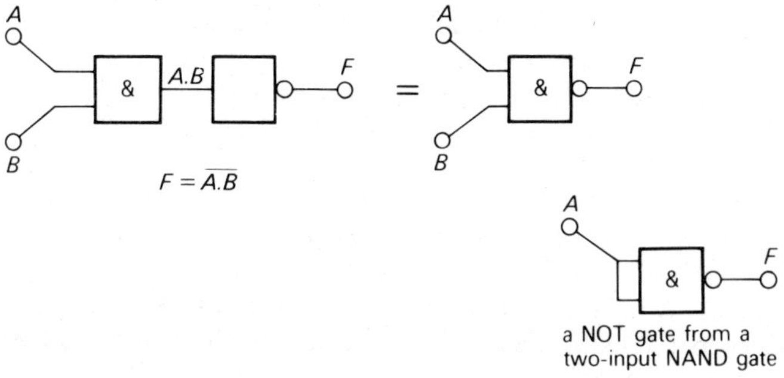

Fig. 6.10 A NAND gate.

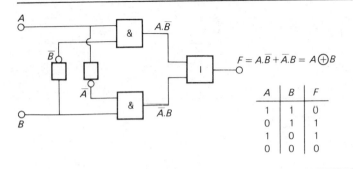

Fig. 6.11   An exclusive OR circuit.

6.6 Linked logic gates

Logic elements AND, OR and NOT can be interconnected to give a wide variety of functions. The exclusive OR circuit for example can be achieved using one OR gate, two AND gates and two NOT gates (Fig. 6.11).

$F = A \oplus B$, i.e. $F = A$ or B but not both.

$\therefore F = A$ and not B or B and not A

$= A \cdot \bar{B} + B \cdot \bar{A}$

An example typical of a situation in industry occurs when a sequence of operations is to be carried out before a machine can be switched on. Such details are as follows:

A machine cannot be started until

1. The start button is pressed.
2. A guard is in position.
3. The speed setting is at its lowest setting.

The machine must stop if the guard is removed or if the machine overloads. The machine must also stop if the **stop** button is pressed.

If positive logic is used it is necessary to ensure that each of the operations outlined applies a positive voltage to the system.

Now the first three conditions are all AND conditions and the gate needed is a three-input AND circuit. The final conditions are essentially NOR conditions. The required logic and the motor circuit are illustrated in Fig. 6.12.

The setting up of logic circuits to achieve a required end result can be produced from the Boolean algebraic expression which in turn has been derived from a series of statements.

Take the example of a drink-vending machine. Let us assume that we have the alternatives of either tea or coffee and we can have either milk

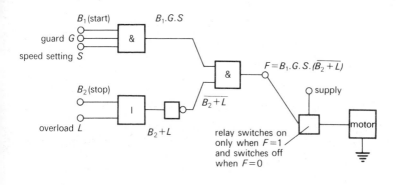

Fig. 6.12   Logic circuit for machine.

and/or sugar with our choice of drink. We are, however, denied having either tea or coffee on their own. Furthermore we cannot obtain only milk or sugar and a mixed tea and coffee beverage is prohibited.

If Tea is represented by T
    Coffee is represented by C
    Milk is represented by M
    Sugar is represented by S

and obtaining the required drink is F, then F = tea or coffee with, in each case, milk or sugar or both.

exclusive OR

$F = T \cdot (M + S)$	$\oplus$	$C \cdot (M + S)$
Tea and milk *or* sugar		Coffee and milk *or* sugar
or Tea and milk *and* sugar		or Coffee and milk *and* sugar

$T \cdot (M + S)$ can be achieved by one AND gate plus one OR gate
$C \cdot (M + S)$ can be achieved with one additional AND gate

The outputs must now be combined in an exclusive OR circuit. Figure 6.13 shows how this can be achieved and the associated truth table is alongside. Only six allowable alternatives are available. The mechanism operating the machine will only work when an acceptable combination of buttons are pressed, that is when $F = 1$.

6.7 Boolean algebraic rules

Boolean algebra behaves in a somewhat different fashion to normal algebra. Already certain symbols (+ meaning OR and · meaning AND) have different meanings from their normal algebraic counterpart.

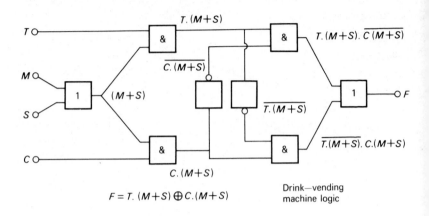

$$F = T.(M+S) \oplus C.(M+S)$$

Drink—vending machine logic

T	C	M	S	F	Remarks
0	0	0	0	0	
1	0	0	0	0	
0	1	0	0	0	
1	1	0	0	0	
0	0	1	0	0	
1	0	1	0	1	Tea and milk
0	1	1	0	1	Coffee and milk
1	1	1	0	0	
0	0	0	1	0	
1	0	0	1	1	Tea and sugar
0	1	0	1	1	Coffee and sugar
1	1	0	1	0	
0	0	1	1	0	
1	0	1	1	1	Tea, milk and sugar
0	1	1	1	1	Coffee milk and sugar
1	1	1	1	0	

Fig. 6.13   Drink vending machine — logic circuit.

Nevertheless a number of algebraic rules apply equally well in both systems; e.g.,

$A \cdot B = B \cdot A$ This is the commutative rule for AND

$A + B = B + A$ This is the commutative rule for OR

$A + (B + C) = (A + B) + C = A + B + C$

$A \cdot (B + C) = A \cdot B + A \cdot C$

$A + 0 = A$

$A \cdot 0 = 0$

$A \cdot 1 = A$

Where things are different a truth table will quickly reveal the true relationship.

Thus $A + A = A$

(There is no meaning to $2A$)

A	A	$A + A = A$
1	1	1
0	0	0 A or $A = A$

$A \cdot A = A$

(There is no meaning to A^2)

A	A	$A \cdot A = A$
1	1	1
0	0	0 A and $A = A$

$A + 1 = 1$

(Anything or 1 is 1)

A	1	$A + 1 = 1$
1	1	1
0	1	1 A or $1 = 1$

$A + \overline{A} = 1$

A	$\overline{A}$	$A + \overline{A} = 1$
1	0	1
0	1	1 A or not $A = 1$

$A \cdot \overline{A} = 0$

A	$\overline{A}$	$A \cdot \overline{A} = 0$
1	0	0
0	1	0 A and not $A = 0$

and $A + (B \cdot C) = (A + B) \cdot (A + C)$ proved by the truth table below:

A	B	C	$B \cdot C$	$A + (B \cdot C)$	$A + B$	$A + C$	$(A + B) \cdot (A + C)$
1	1	1	1	1	1	1	1
0	1	1	1	1	1	1	1
1	0	1	0	1	1	1	1
0	0	1	0	0	0	1	0
1	1	0	0	1	1	1	1
0	1	0	0	0	1	0	0
1	0	0	0	1	1	1	1
0	0	0	0	0	0	0	0

$$\llcorner\text{------- compare -------}\lrcorner$$

Two very important relationships

$$\overline{A \cdot B \cdot C} = \overline{A} + \overline{B} + \overline{C}$$

$$\overline{A + B + C} = \overline{A} \cdot \overline{B} \cdot \overline{C}$$

known as De Morgan's Rules can be verified by truth tables.

The negated expression, e.g. $\overline{A}$, is called the complement of that expression.

Thus $\bar{A}$ is the complement of A.

In many logic gates the complement is available as well as the original expression.

It is possible to achieve a particular logic situation in a variety of ways and obviously some methods use less gates than others. It is often possible to cut down on the number of units required by combining logic units or by producing a different circuit to solve a problem. There are several techniques for minimisation but in simple cases a reduction in the number of gates used can frequently be achieved by inspection. One technique to achieve minimisation is called Karnaugh mapping.

6.8 Fluidic devices (digital)

Logic 'gates' need not necessarily be electronic. It is possible to provide similar functions using a gas or liquid as the 'current' flow. Such components are referred to as 'fluidic' devices. The basic principle relies on the 'Coanda effect'. When a jet of air leaves a nozzle (Fig. 6.14) the air stream outside the nozzle flows in a uniform pattern. When a stationary surface is placed near one side of the jet it produces a viscous drag and layers near the boundary slow down the air flow, thus causing the air to reverse in direction. A turbulence is generated which affects the rest of the air stream causing it to swing towards the boundary surface. This is in effect a switching action. In some fluidic devices this switching action is produced by the physical movement of the boundary surface (usually a thin diaphragm). In others the switching action is achieved by a control air jet at right angles to the main stream virtually blowing it into a different channel (Fig. 6.15). There is a great deal of similarity between the action of the fluidic gate and the electronic gate. In each case for example a supply is needed, air pressure in one instance, a voltage (an electrical 'pressure') in the other.

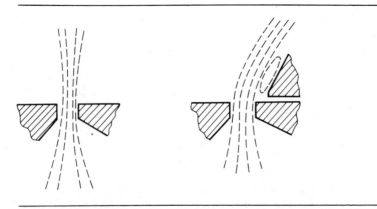

Fig. 6.14   Coanda effect.

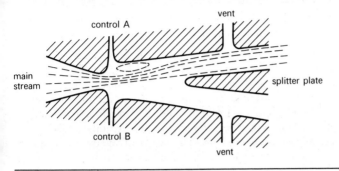

Fig. 6.15   Switching by air jets.

Fluidic OR circuits are essentially two gates in parallel. Fluidic AND circuits are essentially two gates in series. The two-way switch is produced by the device shown in Fig. 6.16 and at the same time this provides an AND output. The main stream of air is deflected into one or other of the exclusive OR channels when a control jet is applied either to A or B. Application of a control jet to A and B causes the main stream to flow in the central channel.

Fluidic logic gates suffer a number of disadvantages:

1. They are relatively slow acting. Each gate may require 2 or 3 ms to operate.
2. They are relatively large — certainly much larger than the equivalent electronic device.

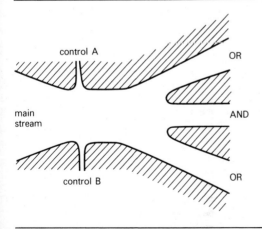

Fig. 6.16   Exclusive OR and AND outputs.

3. They can be affected by environmental conditions, e.g. dust can collect in the nozzles which can either block or divert an air stream.
4. They are much more expensive than electronic units.

On the other hand they are extremely rugged and can operate in certain environments in which electronic devices would fail, e.g. high environmental temperatures and radioactive areas.

Because pneumatic systems are used extensively in the control of machine tools (lathes, milling machines, etc.), there is a tendency to employ fluidic logic in such cases, the air supply is readily available and there is not usually a space problem. The speed of operation, although relatively slow, is usually fast enough for machine tool operation. The high cost, low packing density and low speed makes fluidic devices unsuitable for computers although one or two attempts have been made to produce them.

6.9 Fluidic devices (analogue)

It is possible to amplify a control signal by means of fluidic devices. One form is the vortex amplifier (Fig. 6.17.) Here air enters the main vent at the top and crosses the radial gap and leaves by the central orifice. There is little pressure drop across the device as the air stream merely has to negotiate a right-angled bend. If a further stream of air enters tangentially the main stream is deflected into a spiral path. This produces a pressure gradient in the device and the pressure drop from main input to output is increased. Over a limited input pressure range the output pressure is a linear function of the input. The output pressure change being larger than the input pressure change provides a basis of amplification. Several inputs to the device can be arranged if required.

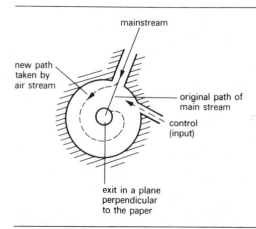

Fig. 6.17 Vortex amplifier.

Summary

Logic statements can be expressed in Boolean algebra. There are four main statements: OR, exclusive OR, AND and NOT. Each can be achieved by a 'gate', usually electronic in design but sometimes using a fluid rather than electric current.

Logic units may be linked to form NOR and NAND gates and combinations of units may be set up from the Boolean algebra. The logic expression can be shown in tabular form known as truth tables.

Questions

Boolean algebra does not obey the same laws as normal algebra. Find the result of $(A + B + C) \cdot (A + B)$.

How does positive logic vary from negative logic?

Venn diagrams are sometimes employed in logic. What are they? Use them to show the difference between $A + B$ and $A \oplus B$.

Digital signals can be expressed in binary so that the number 6, for example, is 110. A number of other codes exist for transforming denary numbers (units, tens, hundreds, etc.) into binary (ones and zeros). One such code is called Binary Coded Decimal or B.C.D. Find out about it and then express the numbers 9001; 909; and 87 in this code.

Examples

1. Simplify the following Boolean expression as far as possible and sketch the associated logic diagram using AND, OR and NOT gates.

$$F = A \cdot \bar{B} + A \cdot (\bar{A} + \bar{B}A)$$

Ans. $A \cdot \bar{B}$

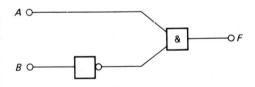

2. Show by truth tables that the following identities hold:

(a) $(\overline{A \cdot B \cdot C})$ $= \bar{A} + \bar{B} + \bar{C}$

(b) $(\overline{A + B + C}) = \bar{A} \cdot \bar{B} \cdot \bar{C}$

(c) $(A + \bar{A}B)$ $= A + B$

(d) $(\bar{A} + AB)$ $= \bar{A} + B$

3. Simplify the following and reproduce as a logic circuit using

(i) AND OR and NOT gates

(ii) NAND gates only

$F = A \cdot B \cdot C + \bar{A} \cdot B \cdot C + A \cdot \bar{B} \cdot C + A \cdot \bar{B} \cdot \bar{C}$

Ans. $F = A \cdot \bar{B} + B \cdot C$

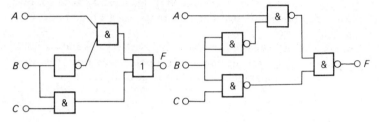

(Belfast, 1975)

4. Sketch a three-input diode transistor logic NOR logic gate. Describe the action of the circuit.

Sketch the Logic Symbol for the gate and the Truth Table for all combinations of input signal.

A	B	C	F
1	1	1	0
0	1	1	0
1	0	1	0
0	0	1	0
1	1	0	0
0	1	0	0
1	0	0	0
0	0	0	1

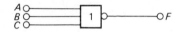

(Norfolk, 1975)

5. Sketch a combinational logic circuit which uses OR, AND, NOT gates to represent the following function

$Z = (\bar{A} + B) \oplus (A + \bar{B})$

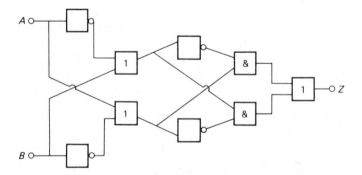

6. In order that a machine may start one of two start switches must be pressed and the safety shield must be in position.

The machine will stop if the safety shield is disturbed, or if one of two stop buttons is pressed.

Draw a logic circuit using AND, OR and NOT gates to control the machine.

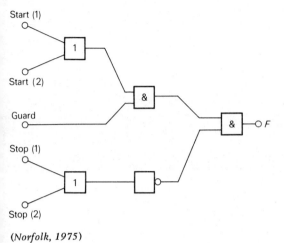

(*Norfolk, 1975*)

Chapter 7

Transducers

7.1 Introduction

Originally transducers were considered to be devices which translated energy from one form to another. They were moreover reversible in their action and could transform the energy either way. Nowadays the term is used rather more loosely and the following definition will be employed in the text.

7.2 Definition

Transducers are devices for producing one variable quantity in terms of another where the two variables are in different energy systems. For example, the aneroid barometer is essentially a transducer where changes in air pressure on a sensitive bellows (a fluid system) results in the rotation of a pointer (a mechanical system). A pressure variable is transformed into a torsional variable. If the translation is mathematically linear then the change in angular position is directly proportional to the change in pressure, i.e. $\theta \propto$ pressure.

Many transducers measure non-electrical quantities which are then translated to electrical quantities. The reason for doing this is because electrical quantities can be measured accurately usually without recourse to elaborate equipment and electrical signals can be transmitted over distances usually more easily than non-electrical.

7.3 An example

A strain gauge is an example of such a transducer. Strain is the ratio of the extension of a sample of material over its original length when that sample is subjected to stress. Often the extension is very small and if a direct measurement is needed then some optical or mechanical amplification of the extension is needed. This involves the use of an expensive piece of equipment (the extensometer) which must be placed on the material undergoing tests and the person making the measurement must be situated close to the apparatus.

A strain gauge is essentially a length of thin nichrome or cupronickel

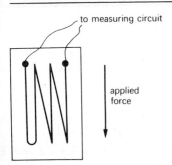

Fig. 7.1   Strain gauge.

wire arranged in a 'W' formation which is cemented to the sample under-going stress (see Fig. 7.1). The strain gauge's resistance changes as the wire is subjected to stress at the same time as the material being tested. The change in resistance (usually measured on a Wheatstone bridge) is an indication of the strain. The bridge may be situated some distance from the gauge and in certain circumstances the strain gauge may be the only means of making the measurement (e.g. measurements on the wing of an aircraft in flight). It is essentially an indirect method of measurement. Sensitive gauges produce a fairly large change in resistance for a given strain but suffer from a major defect in their temperature coefficient of resistance which may also produce changes in resistance. This defect is usually overcome by the use of a second identical strain gauge placed near the first but not subject to any stress. On the assumption that the temperature of the unstressed gauge is the same as the stressed gauge, resistance changes due to temperature changes cancel out (see Fig. 7.2).

It is necessary to calibrate the strain gauge or at least know from the manufacturer of the gauge the change in resistance for a given strain. This information is called the Gauge Factor.

$$\text{The gauge factor} = \frac{\text{Percentage change in resistance}}{\text{Percentage strain}}$$

$$= \frac{100 \times \delta R/R}{100 \times \delta l/l}$$

A typical gauge factor is 2.2. Hence percentage strain = percentage change in $R \div 2.2$.

If the strain is non-static, i.e. is dynamic or changing, then the method of measurement often employed is to balance the Wheatstone bridge initially and then to record the 'out of balance' voltage across the galvano-meter terminals on an oscilloscope. The strain gauge is essentially an electrical method of measuring very small displacements.

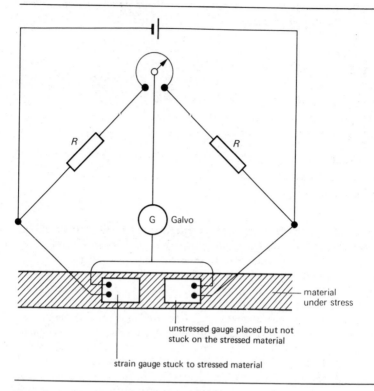

material
under stress

unstressed gauge placed but not
stuck on the stressed material

strain gauge stuck to stressed material

Fig. 7.2   Strain measuring circuit.

7.4 Measurement of the rate of change of displacement d*s*/d*t*

The quantity d*s*/d*t* is a velocity and the method of measurement employed
depends to a large extent upon the magnitude of the velocity. We can
measure the time taken to travel a given distance and this will then give the
average velocity over the given distance. We might wish to know the speed
of a rifle bullet for example or the speed of a car. Obviously these are two
quite different problems, the method employed in the former, where
speeds around 600 m/s are expected, would not be suitable for vehicle
speeds nearer 30 m/s. Moreover the problem of the rifle bullet is very
much a 'one-off' situation, whereas often in the car we wish to know the
speed at any instant – a continuous monitoring of the speed.

With the first measurement of speed, the bullet may be caused to pass
between a pair of light sources and photoelectric cells a fixed distance
apart as shown in Fig. 7.3. This will generate two pulses of electric current
or voltage which after suitable amplification can be applied to one beam of
a double-beam oscilloscope. The first pulse can then trigger off a time base
which causes the spot on the screen to move across the screen at high

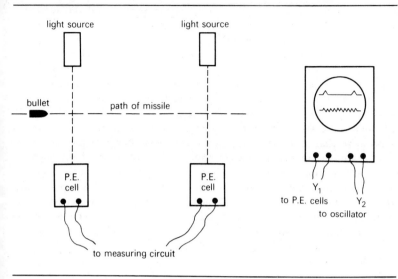

Fig. 7.3   Measurement of a high velocity.

speed. The other beam can be connected to a crystal controlled oscillator operating at precisely 10 kHz. The response is photographed. Now by counting the number of complete cycles on one beam between the 'start' pulse and the 'stop' pulse it is easy to calculate the required time interval as each cycle of the oscillations corresponds to 0.1 ms. Hence if the photo-electric cells are 2.5 m apart and the number of cycles counted on the oscilloscope is 45 then the average velocity of the rifle bullet over this distance is $2.5/45 \times 0.1 \times 10^{-3} = 550$ m/s.

It is nowadays fairly easy to count the number of cycles which have occurred over a short interval by electronic means. If the oscillations occur at 1 MHz it is possible to measure the time interval to ± 4 μs, allowing an accuracy of two orders beyond that indicated by the oscilloscope method (this method is also described in Chapter 4).

The second problem involves the measurement of a much smaller speed but this time a continuous measurement is usually required. The speed of the car is directly related to the rotational speed of the wheels. If the revolutions per minute are measured then knowing the road wheel diameter we have the speed of the vehicle. The rotational speed is measured by coupling a small d.c. generator to the wheels (often via a flexible drive and a gear box). The e.m.f. generated by a d.c. generator is directly proportional to the rotational speed if the field flux is maintained constant. Therefore by using a small permanent magnet to generate the magnetic field the e.m.f. from the rotating armature may be measured by a high resistance voltmeter and the indication is proportional to the speed of the vehicle. Such a device is referred to as a tachogenerator. Variations of the device exist where the permanent magnet rotates and the armature (where the e.m.f. is developed) remains stationary.

Remote indications of speed may also be achieved by using the 'Doppler' effect. This is employed sometimes by the police to detect motorists who are exceeding the speed limit. It consists basically of a very high frequency radio transmitter and receiver. The transmitter produces a continuous output of beamed high frequency oscillations which are reflected by objects (especially metal objects) in the path of the beam. If the objects are stationary the frequency of the reflected radio waves is identical to that sent out. If the object is moving then a frequency shift is produced in the reflected waves, a reduction of frequency if the object is moving away from the transmitter and a frequency increase if the object is moving towards the transmitter. The change in frequency is directly related to the velocity of the object producing the reflection. It is relatively easy to measure the change in frequency between the transmitted and received radio waves.

7.5 Measurement of angular velocity

Three main methods employing transducers are available for measuring the rotation speed of a shaft.

1. Use of the tachogenerator.
2. Use of a stroboflash.
3. Use of the 'toothed' wheel generator.

The first is largely self explanatory. A small d.c. generator has its rotational armature (or field) brought out to a spindle which is in turn placed on the end of the shaft whose speed is to be measured. The shaft causes the spindle to rotate which produces an e.m.f. which is measured on a suitable meter. The e.m.f. is directly related to the rotational speed.

Although this method of angular velocity measurement is straightforward it tends not to be very accurate. Slip often occurs between the shaft and the spindle of the tachometer, resulting in a false reading.

The second method overcomes this difficulty since no contact is made between the measuring device and the rotating shaft.

The flashing speed of a high luminous intensity lamp is controlled by a variable frequency electronic oscillator. When the number of flashes per second correspond with the number of revolutions per second of the shaft the shaft appears stationary. The stroboflash is arranged to illuminate the shaft and usually a mark is placed on the shaft to help the user of the instrument to spot when the shaft appears not to move. It is possible to obtain false readings since flashing speeds which are sub-multiples of the actual speed also produce stationary images. The way of ensuring that the correct speed is being recorded is to increase the flashing speed. If the flashes occur at twice the frequency of the rotational velocity of the shaft then the shaft will have moved only a half revolution between flashes. The

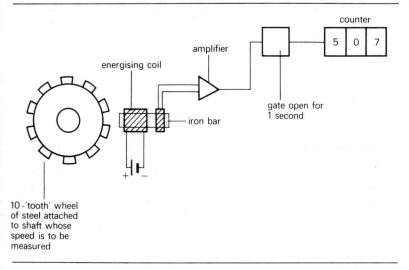

Fig. 7.4   Principle of one form of revolution counter.

mark on the shaft will then appear in two places in synchronism with the flashes. The observer has then only to halve the frequency of the flashes to ensure that he has the right flashing frequency.

If the flashes occur at either a slightly faster rate or a slower rate the shaft will appear to move round very slowly, either clockwise or anti-clockwise.

It is possible to observe quite small variations in rotational velocities by this method.

The third method employs a toothed wheel fixed to the shaft. This wheel is made of ferrous material (see Fig. 7.4). As the wheel rotates the reduction in air gap causes the flux linked by the coil to change and a series of pulses is generated in the pick-up coil. The pulses are counted over a period of perhaps 1 second by a digital counter and knowing the number of teeth on the wheel provides a count of the number of revolutions which have occurred in the second.

7.6 Measurement of acceleration

Since acceleration is the rate of change of velocity any system which produces an e.m.f. proportional to speed can be easily adapted to measure acceleration. The velocity e.m.f. is simply applied to a differentiating circuit (capacitor in series with a resistance) and the voltage across the resistance is approximately proportional to the acceleration (the smaller the time constant the more accurate is the result).

An alternative method is to use the fact that since force F = mass x acceleration then acceleration is directly proportional to the force produced on a given mass. The force may be measured by the extension of a spring and the extension of the spring in turn can be measured by a number of means, if need be by purely mechanical methods. Strictly speaking if the extension is measured by mechanical means a transducer is not involved — there is no change of energy system. Devices which measure acceleration are termed accelerometers.

7.7 Pressure transducers

Two devices worthy of mention are the microphone and the loudspeaker. In the former we are changing from a fluid system (air pressure) to an electrical system. In the second the transducer transforms electrical signals into changes of air pressure. The variation of current in the coil of the speaker causes the lateral force on the coil to vary and the coil moves against the action of a spring. The coil is attached to a stiff cone which in turn causes changes in air pressure and generates sound. This has already been referred to in Chapter 4, Fig. 4.13.

The reverse action — air pressure causing some electrical disturbance — as occurs in the microphone can be achieved in a variety of ways. In one case a flat rectangular crystal of quartz is clamped at one end. When the other end is subjected to a varying force a small e.m.f. is generated across the two faces of the crystal, the polarity and magnitude depending upon the direction and strength of the force applied (see next section).

Piezo-electric effect

Certain crystals which are electrical insulators will produce charges on their surface when they are mechanically deformed. This property is called the piezo-electric effect. Many such crystals show in addition a pyro-electric effect. This is the production of electric charge by heat.

Quartz is commonly used as a piezo-electric material. A piezo-electric coefficient can be expressed in terms of the ratio of the mechanical strain and the electric field it produces on a crystal. The piezo-electric effect works either way; that is, an electric field can produce a mechanical strain in a crystal.

The piezo-electric coefficient $= \dfrac{\text{change of mechanical strain}}{\text{change of electric field}}$ and since strain is non-dimensional $\left(= \dfrac{\text{extension}}{\text{original length}} \right)$ and electric field is expressed in volts/metre the piezo coefficient is measured in metres/volt.

Quartz has a coefficient of around 2×10^{-12} m/V.

A number of other materials — barium titanate and lead zirconate titanate — have a significantly larger coefficient.

The piezo-electric effect provides a useful means of measuring

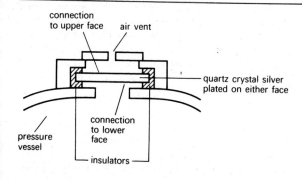

Fig. 7.5 Pressure transducer.

pressures. A typical arrangement is shown in Fig. 7.5. The pressure inside the vessel to which the transducer is fitted causes a deformation of the crystal. This in turn produces an e.m.f. which can be measured. It is necessary to amplify the e.m.f. and the amplifier must normally be a high gain d.c. amplifier with a high input impedance. This type of transducer can respond to very rapid changes in pressure and it is therefore extremely useful in measuring pressure changes inside the cylinders of internal combustion engines.

7.8 Sensors

Sensors are essentially transducers where the position of an object is related to some output quantity — usually an electrical output.

The linear or angular position of the object can be measured by the use of a simple potentiometer.

Take for example the linear potentiometer shown in Fig. 7.6. If a steady voltage is maintained across the ends a, b, of the resistor the voltage

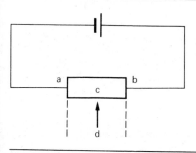

Fig. 7.6 Linear potentiometer.

appearing between a and c is directly proportional to the distance of the arm d from one end. If this arm is attached to the object whose position is required then the output voltage is proportional to distance of the object from some reference. By bending the resistor into the shape of a circle the angular position of the arm is directly proportional to the voltage appearing between one end of the resistor and the wiper arm. For example, if we maintain precisely 36 V across the ends of the resistor then every volt corresponds to an angle of 10°. Potentiometers suffer from one major disadvantage, namely the wear caused by the continual wiping action of the contact, which may in time cause an electrical breakdown.

One way of avoiding this is to use a differential transformer (Fig. 7.7).

An alternating voltage is maintained across the large winding and the outputs of the two secondaries are in opposition. When the core is in the mid-position the coupling from the main primary windings to the secondaries is the same. Two equal voltages appear across the two secondaries but due to the mode of connection the total output voltage is zero.

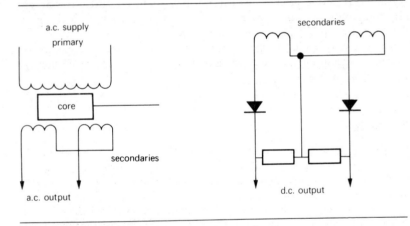

Fig. 7.7 Differential transformer.

Fig. 7.8 Differential transformer arranged for a d.c. output.

Displacement of the core from its central position disturbs the voltage balance and the secondary voltage is no longer zero. If two rectifiers are employed as shown in Fig. 7.8, the d.c. output voltage varies in magnitude and polarity with the position of the core. If the core is placed on a spindle then the sensor detects angular position.

Sometimes the reverse procedure is needed; that is, the position of a device is an indication of some quantity. The pointer of a moving coil ammeter is a simple and obvious example where the current through the coil produces a torque acting against the action of a control spring. The system attains a steady angular position dependent upon the current

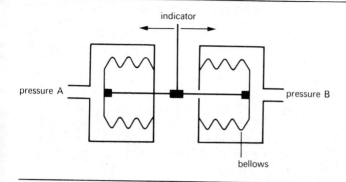

Fig. 7.9 Differential air bellows.

flowing. The angular position is shown by a pointer attached to the coil.

One form of fluid sensor has already been mentioned at the beginning of this chapter — the aneroid barometer. A modification of this is a differential bellows (Fig. 7.9). The position of the pointer arm is dependent upon the difference between the two pressures A and B.

Another form of sensor which is used extensively in pneumatic control systems is the flapper and nozzle (Fig. 7.10). Air at a steady pressure is passed through the restriction at a, passes down the tube, some continuing on to the nozzle, the remainder to the output. Providing the nozzle is unobstructed the air pressure between it and the restriction is low and the output pressure P_0 is small. If the flapper is placed near the nozzle obstructing the air flow the pressure in the tube increases and the output pressure increases.

In practice the flapper is connected to some position-sensing element so that the output pressure is a function of the position of the flapper.

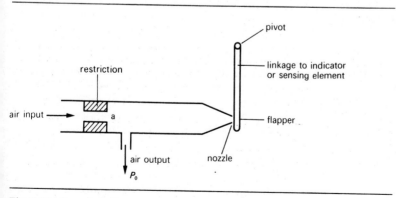

Fig. 7.10 Pneumatic flapper and nozzle.

7.9 Other transducers and means of measurement

A number of other transducers have already been referred to. The measurement of temperature by means of a change in resistance or the generation of an e.m.f. from a thermocouple is one example.

Thickness of plating on sheet steel can be found by a magnetic process. The plating forms a non-ferrous gap between the detecting head and the steel and the flux produced is affected by the thickness of this gap.

Distances of objects from an observer can be measured by noting the time taken for an emitted sound pulse from a source to be reflected back to the source. This is the principle employed in depth sounding at sea where a continuous record of the profile of the sea bed is obtained by sending out a continuous series of pulses.

Transducers cover a very wide range of devices. Increasingly use is being made of semiconducting materials. The reverse current in a pn junction for example is dependent upon the light falling on it. Consequently this can form the basis of light measurement. In general transducers may be represented as 'black boxes' with an input and an output. The relationship between output and input is the transfer function of the device and it is more than probable that the relationship is non-linear.

7.10 Summary

Transducers are devices which translate energy from one form to another. They often translate non-electrical quantities into electrical. They can be used to measure, for example, temperature, pressure, rotational speed and acceleration.

Sensors are merely a special form of transducer used for determining the position of an object and relating this position to some parameter (typically voltage or pressure).

Questions

1. Four identical strain gauges are mounted on a beam; two are in tension and two in compression when the beam is deflected. The gauges are connected in the form of a bridge, to give the maximum output. Show how they are connected. The bridge supply is 5 V and the gauge factor is 2.2. If the strain in the beam is 500×10^{-6}, find the output across the galvanometer terminals if the bridge was initially balanced before the beam was stressed.
Ans. 5.5 mV

2. Draw a diagram of a typical accelerometer and explain how it operates.

3. How does the altimeter on an aircraft function?

Chapter 8

Signals and data transmission

8.1 Information

In engineering systems it is necessary to collect, store and transmit information of one sort or another. The information may be simply of the form that a particular condition has been attained, such as the required temperature of a furnace has been reached. It might be in the form of some numerical value — for example, that so many components in a production process have been manufactured. It might be a much more complicated and lengthy item of information such as a report on the quality of a particular steel. Complex information is normally expressed in words, numbers and symbols in a written report and stored on paper in a filing system. It is usual to give it some form of address or reference number so that it can be easily found when needed. Large quantities of information are stored in books or pamphlets placed on library shelves. A method of address is used (such as the Dewey classification system) so that books can be easily located.

The problem nowadays is storing masses of information in a reasonably accessible form and in a reasonably economic manner. Frequently the information is reduced in physical size by using microfilm techniques. Occasionally the information is stored within a computer, again with a coded address so that it can be traced quickly when needed.

Often information is presented in a standard form as with a card index system so that a large amount of information can be seen at a glance especially by a trained observer. Sometimes the information is stored by punching holes in cards or tape or by magnetising metallised tapes or discs.

The simplest form of information can be presented visibly by an indicator lamp which glows when a required condition has been reached, e.g. a green lamp might glow when the required temperature has been reached in a furnace. Numerical information can be presented by a counter display, e.g. the mileage indicator of a car. More complex information is displayed in words, numbers and symbols or by a diagram and photograph. Information stored in a computer on magnetic tape is normally fed into a teletype and displayed as typewritten characters.

8.2 Transmission of information

Information is often required to be transmitted some distance from the source. The method used to transmit the information depends upon how complex the information is, the distance over which it is to be transmitted and the speed at which the information is needed.

In the example just given the furnace might be located only a few metres from the control panel where the indicator lamp is situated. The connection from the temperature-sensing element at the furnace to the control panel is direct by means of conducting wires. There is no time delay in transmitting the information that the required temperature has been reached; information from a computer to a teletype might have to be transmitted several kilometres. Although each item of information is transmitted virtually instantaneously, it takes a significant time for a quantity of information to be presented by the teletype.

On occasions when there is no urgency in passing information from one place to another it is possible to use the postal system and convey in a book or written report a considerable quantity of information with a delay of possibly several days (especially using second-class mail). Where it is necessary to have an exchange of information delays of several days may be unacceptable and one then has to use a telex system or the telephone.

Much depends upon the complexity of the information and the speed at which it is needed. Generally speaking the greater the speed of transmission the more sophisticated is the method of transmission and the greater the cost involved, although the telephone in this country is still a surprisingly cheap mode of communication in spite of recent increases.

8.3 Basic requirements of a transmission system

In communication systems the information required has to be obtained, coded, addressed, transmitted, received and then decoded by the recipient of the information.

The sequence of the postal mode of communication as an example is as follows:

1. Information obtained.
   A technical report on the testing of a sample of steel for instance is received from the materials and chemical laboratories.
2. Information coded.
   The results of the test are expressed in words and symbols and put down on paper in a written report.
3. Information addressed.
   The report is placed in an envelope addressed to the person requiring the information.
4. Information transmitted.
   The report is posted and is now in the hands of the Post Office with a delay of 1 or 2 days (or more!)
5. Information received.
   The recipient receives the envelope and its contents.

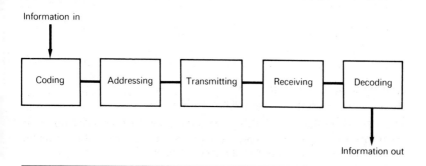

Fig. 8.1 Basic requirements of a transmission system.

6. Information decoded.
   Recipient reads the report and gleans the information (see Fig. 8.1).

Other methods of transmission follow a similar pattern but the coding and decoding processes are often automatic and the transmission time is very short.

It should be pointed out that errors can occur at many stages in the communication process. The information may be wrongly coded; for example, the incorrect temperature may be recorded, the wrong address may be used and the information may be lost or may go astray, or the recipient may misread the information because of ambiguity or a typing error. Seldom is the complete system perfect but often transmission errors are insignificant. An odd word is lost perhaps in a telephone conversation or a wrong spelling occurs in a telex message but the sense of the information is unaltered. One must be cautious however and realise that the omission of a single word can sometimes drastically change the sense of a message. 'I will not be able to get to the meeting' is quite a different piece of information if the word 'not' is removed.

8.4 Analysis of signals

A signal may be simply a switching action (ON or OFF) which results, for example, in a lamp glowing. Such a signal is essentially a step change and is reminiscent of a binary signal.

More complicated signals can be sent by a series of binary (ON/OFF) changes.

The morse code of dots and dashes uses this principle whereby a series of two level signals constitute first letters and then words. The rate of signalling is relatively slow if each element of signal is to be sent by essentially a manual process (see Fig. 8.2). The punched tape is certainly a much speedier process since each letter requires only one combination of holes

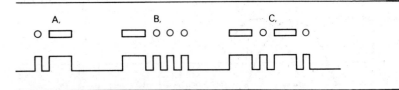

Fig. 8.2   Morse code as an example of a two-level signal.

in the tape (see Fig. 8.3*a*). Five-hole tape can produce 2^5 or 32 different combinations, eight-hole tape can produce 2^8 or 256 different combinations and therefore 256 different characters.

Much of the data transmission may be conveyed by means of binary signals over short distances as with a computer or over relatively long distances via telephone lines. Such signals are referred to as digital signals.

Many signals however are not of this form but are essentially variations in amplitude of some variable quantity such as voltage. These are called analogue signals. The simplest form of analogue signal is the sine wave which follows a well prescribed equation: $y = A \sin \omega t$. A is the amplitude of the signal and $\omega = 2\pi x$ frequency f of the signal. f is measured in Hertz (Hz).

Sine waves of variation in air pressure can be heard if the frequency lies in the range 50 Hz–20 kHz approximately.

The loudness of the note which is heard depends upon the amplitude A; the pitch of the note depends upon the frequency f.

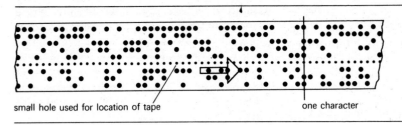

small hole used for location of tape   one character

Fig. 8.3*a*   Punched tape as used in a computer.

```
10 DIM M(1000)
20 MAT M=CON
30 RANDOMIZE
40 PRINT "THE EXPECTED %AGE FAILURE RATE AT EACH STAGE IS";
50 INPUT F
60 LET F=F/100
70 PRINT "THE INITIAL QUANTITY (LESS THAN 1000) IS";
80 INPUT N1
90 LET N0=N1
91 PRINT "THE NUMBER OF FAILURES";
```

Fig. 8.3*b*   The data contained in the punched tape.

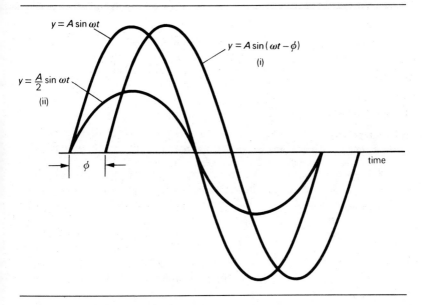

Fig. 8.4 Variation of phase (i) and amplitude (ii) of a sine wave.

Sine waves can also be moved in position by introducing a phase angle ϕ, thus $y = A \sin (\omega t - \phi)$. (See Fig. 8.4.)

ϕ can obviously vary between 0 and 2π radians and the shift in time depends upon whether ϕ is positive (phase advance) or negative (phase retard). Information can be conveyed by altering A, ω (or f) or ϕ.

Complex information (such as speech) consists of several sine waves of varying amplitude, frequency and phase being added together. The result no longer resembles a simple sine wave (Fig. 8.5) but any complex wave of this sort can be split up into its component sine waves. When a single sine wave is transmitted the signal is referred to as a single tone. If sine waves having a frequency n times a particular frequency (n being an integer) are transmitted these are referred to as harmonics.

The lowest frequency is called the fundamental ($n = 1$) and when $n = 2$ the signal is referred to as the second harmonic.

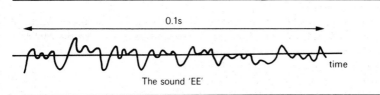

Fig. 8.5 The sound 'EE'.

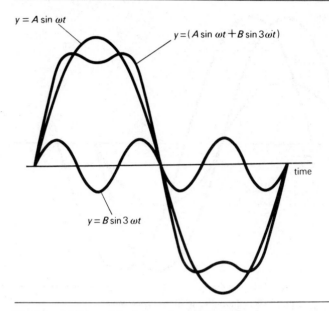

Fig. 8.6   Addition of fundamental and third harmonic.

The effect of adding a third harmonic ($n = 3$) to the fundamental is shown in Fig. 8.6.

Speech frequencies have a frequency range of about 50 Hz–4 kHz and this is the approximate frequency range of the telephone system. The term bandwidth is also indicative of the frequency range and the speech or audio frequency range can be said to require a bandwidth of about 4 kHz.

Good quality music requires a larger bandwidth and television signals require an even greater one – typically 5 MHz. The more information sent in a given time the greater is the bandwidth required.

8.5 Amplitude modulation

It is sometimes necessary to shift bodily a signal from its original frequency band to a new one. This can be readily achieved by 'modulating' one sine wave with another.

The simplest method is called amplitude modulation and consists of varying the amplitude of a high frequency sine wave – called the carrier – by the signal which is to be transmitted.

Let the signal be a single tone sine wave

$$v = v_s \sin \omega_s t$$

$\omega_s = 2\pi \times f_s$ the signal frequency.

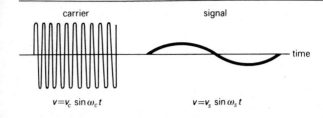

$v = v_c \sin \omega_c t$

$v = v_s \sin \omega_s t$

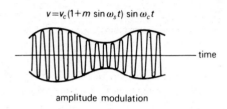

amplitude modulation

Fig. 8.7   Amplitude modulation.

Let the carrier be represented by

$v = v_c \sin \omega_c t$

$\omega_c = 2\pi f_c$ the carrier frequency. f_c is usually very much greater than f_s.
    The effect of modulating the carrier is shown in Fig. 8.7.
    The equation of the modulated sine wave is

$v = (v_c + v_s \sin \omega_s t) \sin \omega_c t.$

A factor m is often used called the modulation index and is the ratio v_s/v_c.
Normally m is less than 1.0.
    The expression may be rewritten $v = v_c (1 + m \sin \omega_s t) \sin \omega_c t.$
    On multiplying out

$v = v_c \sin \omega_c t + m \, v_c \sin \omega_c t \sin \omega_s t$

now $\cos(A - B) = \cos A \cos B + \sin A \sin B$
and $\cos(A + B) = \cos A \cos B - \sin A \sin B$
Hence $\cos(A - B) - \cos(A + B) = 2 \sin A \sin B$
or $\sin A \sin B = \tfrac{1}{2}[(\cos(A - B) - \cos(A + B))]$

It follows that $\sin \omega_c t \sin \omega_s t$ can be expressed as

$\tfrac{1}{2} \cos(\omega_c - \omega_s)t - \tfrac{1}{2} \cos(\omega_c + \omega_s)t.$

The modulated wave is therefore given by the following:

$v_c \sin \omega_c t(1 + m \sin _s t)$

$$= v_c \sin \omega_c t + \tfrac{1}{2} m v_c \cos(\omega_c - \omega_s)t - \tfrac{1}{2} m v_c \cos(\omega_c + \omega_s)t$$

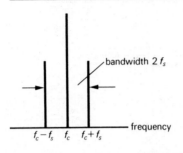

Fig. 8.8   Amplitude modulation frequencies (side bands).

A cosine wave is merely a sine wave phase shifted 90° so that it follows that when amplitude modulating a carrier by a single tone we end up with three frequencies, the carrier, the difference between carrier and signal (this is called the lower side band frequency) and the sum of carrier and signal (the upper side band frequency) (see Fig. 8.8). The bandwidth is now double that of the original but the signal is shifted bodily in the frequency spectrum.

Modulation can be achieved by means of a non-linear device.

If a system has a current/voltage characteristic of the form $i = a + bv + cv^2$, this is referred to as a 'square law' device.

Applying an input voltage v causes an output current i.

If v is a sine wave $A \sin \omega t$ then

$$i = a + bA \sin \omega t + cA^2 \sin^2 \omega t$$

$$= a + bV \sin \omega t + \frac{cA^2}{2} - \frac{cA^2}{2} \cos 2\omega t \qquad \left(\text{since } \sin^2 A = \frac{1 - \cos 2A}{2} \right)$$

and a new frequency double that of the input appears. This is normally referred to as a second harmonic distortion term.

If v is now two sine waves in series

$$A \sin \omega_s t - \text{a signal}$$

and $\quad B \sin \omega_c t - \text{a carrier}$

$$v_{in} = A \sin \omega_s t + B \sin \omega_c t$$

$$i_{out} = a + bA \sin \omega_s t + bB \sin \omega_c t + c(A \sin \omega_s t + B \sin \omega_c t)^2$$

The last term can be expanded to

$$c(A^2 \sin^2 \omega_s t + 2AB \sin \omega_s t \sin \omega_c t + B^2 \sin^2 \omega_c t)$$

The first and last factors

$$A^2 \sin^2 \omega_s t \quad \text{and} \quad B^2 \sin^2 \omega_c t$$

generate second harmonic terms as previously illustrated.

The factor $2AB \sin \omega_s t \sin \omega_c t$ can be written as

$AB \cos(\omega_c - \omega_s)t - AB \cos(\omega_c + \omega_s)t$.

Grouping these frequencies with $bB \sin \omega_c t$ we can see that i_{out} contains the frequencies

$(\omega_c - \omega_s)$ — lower sideband frequency
ω_c — carrier frequency
$(\omega_c + \omega_s)$ — upper sideband frequency

If all the other frequencies in i_{out} are filtered away; only the three terms forming a modulated output remain.

8.6 Multiplexing

The modulation process can be used to transmit a number of signals simultaneously over one pair of wires.

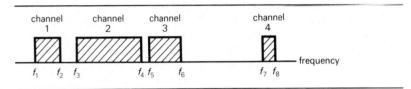

Fig. 8.9   Multiplexing.

The various signals to be transmitted are made to modulate a number of different carriers so that signals now occupy various bandwidths as shown in Fig. 8.9. It is necessary at the receiving end to filter out the required signal from all the others. A process of demodulation is then necessary to remove the now unwanted carrier and retrieve the original signal. The filters which allow a required band of signals to pass on into the next stage of the system are called band pass filters and they effectively stop unwanted signals. The demodulator is basically a rectifier which, in conjunction with a suitable capacitor, rejects the high-frequency carrier. The demodulator circuit is shown in Fig. 8.10. The choice of the value of the

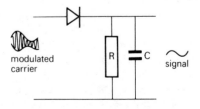

Fig. 8.10   Demodulator circuit.

capacitor C depends on the value of R and the modulation index m. It must not exceed a critical value otherwise distortion occurs.

The rectifier removes the bottom half of the modulated wave and the average value now follows the waveform of the original modulating signal. The high frequency spikes are removed by the capacitor C so that only the required signal is now passed on to the output. Usually an amplifier is needed between the demodulator and the output. The amplifier has to be designed to cope with the frequency of the signal.

A system which uses a number of different carriers is referred to as a carrier system or frequency division multiplex system.

An alternative system is called a time division multiplex system in which signals from various sources are sent essentially in sequence but with an extremely rapid transfer from one signal to the next. This system will not be described.

8.7 Signal delays and signalling speeds

Delays in transmitting signals depend upon the system used. The postal system is one which involves a delay of 1 or 2 days between posting (sending) and delivery (receiving). With electrical systems the transmission time is very short even over quite considerable distances although delays over telephone lines can amount to a few milliseconds. Sometimes the signalling system uses a fluid or gas and the propagation time is significantly greater than the electrical system. Here the signal may be a change in pressure as in a pneumatic system. A typical time of travel in such a system is around 300 m/s compared with 25,000 km/s in a telephone circuit. The speed at which information is to be transmitted is governed by the system employed and the type of information. Speech requires a bandwidth of 4 kHz and the speed with which the signal is transmitted depends upon the speed at which a person speaks. It is possible to increase the signalling speed by recording the speech on magnetic tape at a low speed (say a tape speed of 5 cm/s) and transmitting at a higher speed (say 20 cm/s). The signal will be transmitted at four times the speed but the bandwidth required is now 4 x 4 kHz = 16 kHz. The information contained in a picture can be sent either in a few milliseconds or over several minutes. In the former case a large bandwidth (4 MHz) would be needed, in the latter case only a relatively small bandwidth (4 kHz).

The method employed to transmit a picture is essentially by 'scanning' the picture in a series of lines in much the same way as reading the lines of print in a book. The BBC uses 625 lines in its TV picture scan. Different light tones in a black and white picture are then transformed into voltages of varying amplitudes, perhaps 0.5 V corresponding to 'white' and 0.1 V corresponding to 'black' (see Fig. 8.11). If colour is to be transmitted this is additional information and the bandwidth may have to be still further increased.

The bandwidth required for a signal depends upon the complexity of

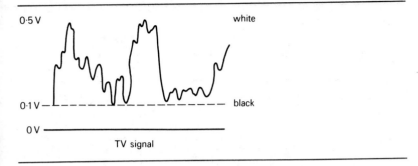

Fig. 8.11 TV signal.

the information and the speed at which it is to be transmitted. Increase either, and the bandwidth needed goes up.

8.8 Signal distortion — noise

Distortion of a signal can cause a number of different things to happen. If A in Fig. 8.12 represents an amplifier in a transmission system the various forms of distortion will be clearly seen — amplitude or frequency distortion, phase distortion and harmonic distortion.

The causes of distortion could be due to the transmitting or receiving apparatus but the connecting link between sender and receiver may also be responsible for distortion. One cause of distortion in transmission is the fact that all frequencies do not travel along the transmission line at the same speed and therefore arrive at the receiver at different times. This is essentially phase distortion and has little effect as far as audio signals are concerned but has a significant effect on the quality of received video signals.

With analogue systems there is a lower limit in the strength (size) of signal which can be transmitted. When the signal level is down to microvolts or less the perturbations always present in random electron movement in resistors and transistors present in the transmission system generate very small voltages of the same order of magnitude as the signal itself. This irregular variation consists of spurious spikes of voltage of random height and random position and is called noise (Fig. 8.13). If noise is suitably amplified and applied to a pair of headphones it sounds rather like 'hissing'. It is obvious that if the required signal is of the same order of magnitude as the average noise great difficulty will be experienced in understanding the signal. This is especially so with analogue signals. Methods exist with digital signals which enable the required information to be extracted from a very noisy background — see section 8.10.

The ratio of required signal/unrequired signal is referred to as the

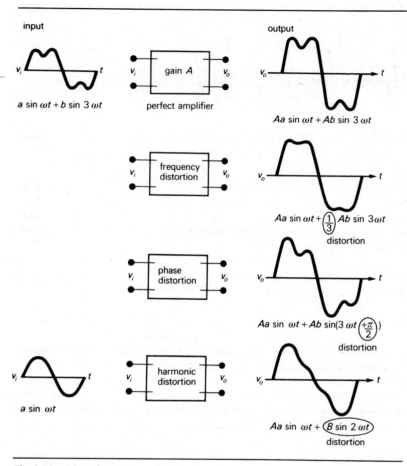

Fig. 8.12   Distortion in an amplifier.

signal/noise ratio and the greater this ratio the more easily is the required signal understood.

The signal/noise ratio frequently employs the decibel system, i.e.

$$\text{signal/noise ratio in decibels (dB)} = 10 \log_{10} \frac{\text{power in signal}}{\text{power in noise}}$$

(This logarithmic scale is also used in acoustics (Chapter 4).)

Noise in an electrical transmission system can be man-made. The interference on a TV picture by the ignition system of a passing car is one example. Such noise may cover a comparatively narrow frequency band but the random noise in resistors and transistors covers a very wide band indeed and can be reduced to negligible proportions only by reducing the random movement of electrons. This can be achieved by operating the

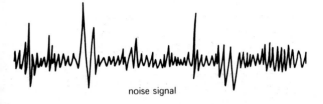

noise signal

Fig. 8.13   Noise signal.

equipment at very low temperatures — approaching absolute zero (−273°C or 0 K). Noise which covers a very wide frequency band with no predictable pattern is called white noise. White noise is mathematically described by saying that the power density over a given bandwidth anywhere in the frequency spectrum is uniform.

Imagine that the noise voltage is applied to a resistor R. The power is given by v^2/R. If the voltages in the noise are limited to those having a frequency in the bandwidth f_1 to f_2 then the noise power is

$$\frac{v^2}{R} f_1 \text{ to } f_2$$

This will always give the same result with white noise whatever centre frequency f_1 and f_2 embrace, i.e. the same value would be obtained if f_1 and f_2 were 4 kHz and 6 kHz or 1 MHz and 1.002 MHz.

Since all frequency bands contain the unwanted noise to the same extent it is advantageous to limit the bandwidth of a transmission system to no more than that required by the signal.

8.9 Signal distortion — reflections

When a signal is sent out the magnitude of the signal gets smaller the further one moves from source. This is called attenuation and in an electrical system is due to the resistance of the conductors carrying the signal and the leakage between them. In a fluid system this is due to a viscous drag by the walls of the tube or pipe containing the fluid. It follows that for long transmission paths only a small amount of the original signal is received; the rest is dissipated in the transmission line or pipe. Considering only the electrical case, the very long transmission line behaves as a dissipative element having some resistive value and virtually all the energy in the signal at the sending end is dissipated in the line. If a shorter line is considered and a resistance equivalent to that of the effective resistance of the very long line is placed at the receiving end, the signal from the sending end will be largely dissipated in the receiving end resistance. This is precisely the conditions required (Fig. 8.14). Such an

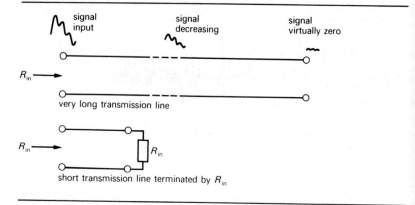

Fig. 8.14   Short transmission line terminated by R_m.

arrangement is known as matching and a matched transmission line, whatever its length, behaves as if it were a very long transmission line. Optimum power transfer conditions then apply. In the Post Office telephone system the matching resistance is 600 Ω for the transmission lines.

A mismatched line does not allow all the power transmitted to be received since reflection occurs at the receiving end and part of the signal is returned to the transmitter. If in turn mismatch occurs here further reflection takes place. The shuttling to and fro of the signal is obviously undesirable because (*a*) further attenuation occurs and (*b*) a series of signals are received rather than a single signal each displaced from the next by the transit time of the signal along the line. This produces distortion. The multiple reflections one obtains in a mirror if one looks into one mirror with a second mirror behind one's head is analagous to mismatch at two ends of a transmission path.

Reflections occur in fluid systems; the so called water hammer is one example, the sound echo is another.

8.10 The binary signal

Many of the shortcomings of analogue signals may be overcome using digital signals and this is especially so when numerical data is being transmitted. With the advent of cheap and small electronic circuits the digital or binary signal is becoming of increasing importance in modern transmission systems.

A binary signal consists of a series of ones (1's) or zeros (0's) sent in a particular sequence which conveys either numbers or letters or symbols. A one (1) is usually regarded as a positive voltage or increased pressure and a zero (0) is usually regarded as a zero voltage or normal pressure. Figure 8.15 shows the number 25 in a decimal system, represented in binary as 11001, and the corresponding signal which would be transmitted. Each 1

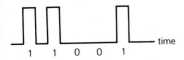

Fig. 8.15 Transmitted binary number 11001 equivalent to decimal number 25.

or 0 is called a binary digit or **bit** and the collection of bits, e.g. 11001, is called a **word** (although in this case it represents a number). Thus in this example the word length is five bits.

The rate at which a series of bits is transmitted is the **bit rate** and in modern systems can be several millions per second. It is necessary to know the bit rate in order to know the number of consecutive zeros in a word. It is generally necessary to know the word length as well.

After transmission the square signals of the transmitted binary word become rounded, i.e. suffer distortion (Fig. 8.16). It is a very simple matter to reconstitute the signal by means of electronic switching circuits because the only decision to be made as each bit arrives is whether it is a 1 or 0. With an analogue signal it would be necessary to know the magnitude of the voltage at each instant. Of course a spurious voltage from a noise source may cause a 1 to be received where a 0 should be.

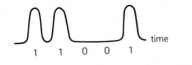

Fig. 8.16 Received binary signal 11001 after some distortion.

It is possible by use of special codes to be able to detect errors of this sort. The principle under which such error detecting codes function is by transmitting words of equal length — say five bits. The first four bits constitute the information in the signal and the final bit is called a **parity** bit. Each word is arranged to have an odd number of ones (1's). If the information in binary terms already contains an odd number of 1's no parity bit is necessary. If however there are an even number of 1's in the message then an extra bit (the party bit) is added (see Fig. 8.17). At the receiving end only words containing an odd number of bits are recorded in the message. It means that errors in transmission are virtually gaps in the information.

This is actually an improvement over a message containing mistakes. For example, if the word **mooms** is received there is obviously a mistake. It could be **rooms, looms, moons, booms, cools, moods, dooms, moors, tools,** etc. It is much more helpful if we know if there are one or more errors in the message and which letter is in error. If the word received is

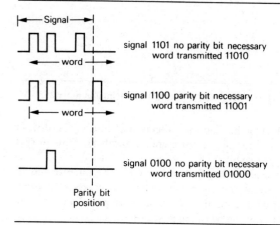

Fig. 8.17   The parity bit.

—ooms then we know that there is only one error and we know the position of the error. The required word is either **booms, dooms, looms** or **rooms**. The other words in the received message will help to identify the required word. The chances of choosing the correct word increase as more of the message is revealed — even when further errors are present, e.g. —ooms, booms, dooms, looms, rooms? but d—ning —ooms must be **dining rooms**. The alternatives do not make sense but in case there is still some doubt the whole message (with further errors) can be correctly interpreted. Hot—l has si—teen b—d—ooms and two d—ning —ooms. In general the greater the length of the message the greater likelihood of being able to correct errors.

This is generally not the case when numbers are transmitted. The number 1—3 can be one of ten numbers 103, 113, 123, etc., and it is unlikely that further information will help. A spelled out number leaves less chance of error, e.g. on— hu—dre— a—— th——e.

It is possible to generate codes in which an error can not only be detected but can also be corrected but this is a subject outside the scope of this book. It will be appreciated that on occasions cancellation of the error detection can occur if noise generates two 1's in a word and such errors may pass undetected. Nevertheless binary signals are far less susceptible to distortion by noise and even when the noise level exceeds the signal it is still possible to extract the information.

This special property of digital or binary transmission is so attractive that it is sometimes advantageous to transform an analogue signal into a digital one before transmission and reconvert the received digital signal to an analogue one. Such converters are termed A to D (analogue to digital) and D to A converters. With the increased use of micro electronics it is relatively easy and cheap to set up circuits which will do this. Furthermore such circuits occupy only a very small amount of space.

It almost goes without saying that non-electrical transmission systems are employed only when relatively simple information is being handled and when the signalling speeds are low. Compressed air is sometimes employed in the control of machine tools. A rather antiquated example of the way compressed air or steam can transmit a message is in the old pianola rolls or the punched sheets used in the fairground steam organ. It is possible to cause machine tools to operate in a prescribed and quite complicated manner by a somewhat similar process.

8.11 Remote position indicators

The information to be transmitted is often of the variety presented by a scale reading on an instrument dial. A pressure scale or temperature may be required to be read remotely. A typical instance is the steam pressure in a boiler in an electricity generating station. The station superintendent wishes to have accurate information readily available at the control panel situated some hundreds of metres from the boiler. A number of possibilities exist, one being the use of a potentiometer in the transmission system. The pointer of the temperature indicating instrument is attached to the wiper arm of a potentiometer (Fig. 8.18). The voltage between one end of the potentiometer and the wiper is proportional to the deflection of the indicating instrument. This voltage can be applied to a pair of transmission lines and a high resistance voltmeter can be connected to the far end. The voltage is now proportional to the temperature. It is of course necessary to rescale the voltmeter in degrees centigrade.

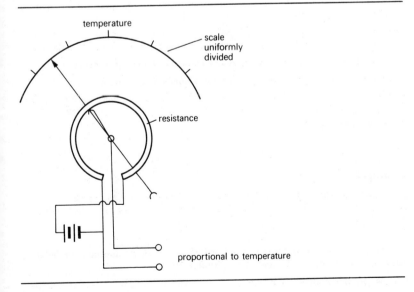

Fig. 8.18   Remote position indicator.

Summary

Information is collected, stored and transmitted. When transmission takes place distortion and delay of the signal may occur. Signals may be in analogue form or digital form. The simplest analogue signal is the sine wave and by amplitude modulation it is possible to shift the signal in the frequency spectrum. Frequency division multiplexing is based on this principle.

Bandwidth requirements increase with the complexity of the information and the signalling speed.

Signal distortion in transmission can be caused by reflections at mismatches and noise. The signal/noise ratio in decibels is $\log_{10}$ Signal power/Noise power.

Digital signals are usually in some form of binary code and the terms bit and bit rate are relevant. It is possible to generate binary signals which have error detecting codes. Errors can often be corrected if sufficient information is transmitted.

Binary signals are less susceptible to noise than analogue signals.

Questions

What is the approximate cost of transmitting a twenty-word message
(*a*) from London to Manchester, (*b*) from London to New York;
(i) by letter, (ii) by cable or telex, (iii) by phone?
If the message contained 2000 words what would be the relative costs?

Comment on the general relationships (if any) between complexity of information, transmission distance and cost of transmission. What are the speeds at which these messages can be transmitted?

The 'address' of a telephone is the phone number. Comment on the way a telephone number is built up. Is there any connection between dialling codes and postal codes?

What are the various sources of noise in an electrical transmission system?

Is it possible to transmit an analogue signal in a fluid system? How?

Before the telephone or electric telegraph was invented how were urgent messages transmitted?

What is a repeater station?

Examples

1. A black and white television picture frame can be split into 625 lines and each line contains 400 bits. If fifty pictures are transmitted per second what is the bandwidth of the signal?
Ans. 6.25 MHz

2. A single frequency of 4 kHz and of amplitude 0.5 V is to amplitude modulate a carrier frequency of 500 kHz and of amplitude 1.5 V.

What is   (i) the modulation index *m*

(ii) the frequencies of the two side bands

(iii) the relative power in the two side bands and the carrier.

Ans. $m = \frac{1}{3}$

   f_1 = 496 kHz

   f_2 = 504 kHz

   ratio 1 : 18

3. The following binary message uses a simple code A = 1 (0001) B = 2 (0010), etc., for letters of the alphabet. The fifth bit in each word is a parity bit. What is the answer?

11100	10011	11100	01011
00010	11100	01000	
11111	11100	01011	
11010	00010	10110	01011?

Ans. TEN

4. There is a limitation to the code in Question 3. What is it? The following message uses the same code and partially overcomes the limitations. What does it say?

01000	10011	01101	01101	10011	00111
01100	11001	01010	10011	01011	01001
00010	00110	01011			
11010	00010	01000	01011		
01010	11111				
00100	01011				
11111	01111	01011	00110	00111	11111	11010	01011.

5. The total power dissipated in a resistor from a signal with noise present is 6 watts. If the signal/noise ratio is 3 dB, what is the noise power? If the noise power is now reduced to 2 mW with the same level of signal power, what is the new signal/noise decibel ratio?

Ans. 3 watts;   35 dB

Index

α, 107
acceleration, 141
accuracy, 9
algorithm, 119
ammeter, 35
Ampere, 3
amplitude modulation, 152
analogue signal, 150
AND statement, 121
angular velocity, 140

β, 108
ballistic galvanometer, 54
bandwidth, 115, 156
binary signal, 149, 160
bit, 161
bomb calorimeter, 80
Boolean algebra, 121
    rules, 127
Bourdon tube, 90
brake, 72
bridge rectifier, 37
brightness, 81
Brinell test, 67

calibration errors, 6
calipers, 61
calorific value, 80
calorimeter, bomb, 80
candela, 4, 81
capacitance measurement, 46
carrier, 152
cathode ray tube, 38
Celsius scale, 4
Centigrade scale, 4
Coanda effect, 130
collector characteristics, 109
common emitter, 107
control torque, 26
current ratio
    α, 107
    β, 108

damping torque, 26
dash pot, 29
decibel, 85, 158
deflecting torque, 26
degree Kelvin, 4
demodulation, 155
dial gauge, 63
differential
    air bellows, 145
    transformer, 144
digital instruments, 56
diode
    logic, 121
    semiconductor, 103
    thermionic, 97
distribution graph, 14
dynamometer
    brake, 72
    instrument, 29
    wattmeter, 31

error graph, 7
errors
    observational, 9
    random, 9
    systematic, 7
    total, 11
exclusive OR statement, 120
extensometer, 65

flapper and nozzle, 145
flowmeter, 91
fluidic devices, 130
fluid measurement, 88
flux
    luminous, 81
    magnetic, 52
fluxmeter, 55
form factor, 38
frequency response, 115
full-wave rectifier, 100
fundamental, 151

Gaussian distribution, 16
grease-spot photometer, 82

half-wave rectifier, 100
Hall effect, 56
hardness, 66
harmonic, 151
histogram, 13
holes, 101
hybrid parameters, 111
hysteresis loop, 54

inductance bridge, 49
information, 147
integrating sphere, 83
interferometry, 69

Kelvin degree, 4
kilogramme, 2

lifetime, 18
Lissajou figure, 40
logic
    statements, 119
    symbols, 124
lumen, 82
luminance, 81
luminous
    flux, 81
    intensity, 81
lux, 82

magnetic flux, 52
manometer, 88
mass, 2, 60
mean, 14
mercury thermometer, 75
metre, 2
micrometer, 61
micron, 5, 60
modulation index, 153
moving coil
    loudspeaker, 87
    meter, 27
moving-iron meter, 29
multiplexing, 155

'n' type semiconductor, 103
NAND gate, 125
noise, 157
NOR gate, 125

normal probability curve, 19
NOT statement, 121

observational error, 9
optical pyrometer, 79
OR statement, 120

'p' type semiconductor, 103
parity bit, 162
phase angle, 151
phon, 86
photometers, 82
photo-voltaic cell, 84
piezo-electric effect, 142
pn junction, 103
pointer instruments, 26
Pitot tube, 93
Poisson distribution, 20
polar diagram, 83
potentiometer, 49
    commercial, 51
    metre wire, 50
pressure, 88
pressure transducer, 142
probability, 18
Prony brake, 72
protractor, 64
pyrometer, 79

quiescent point, 110

radian, 4
random errors, 9
rectification
    bridge, 37
    full-wave, 100
    half-wave, 100
resistance, measurement of, 43
resistance thermometer, 77
reverberation, 86

SI units, 2
scale − vernier, 63
second, 2
secondary standard, 5
semiconductors, 100
sensitivity, 10
sensors, 143
shunt, 35
sidebands, 155
signal delays, 156
signal distortion, 157

significant figures, 21
sinebar, 64
slip gauges, 5
sound, 85
standard deviation, 14
standard error of the mean, 18
standards, 1
steradian, 4
strain gauge, 137
stroboflash, 140
sub-standard, 5
substitution method, 44
surface finish, 69
systematic error, 7

tachometer, 140
Taly-surf, 69
thermionic diode, 97
thermistor, 78
thermocouple, 77
thermometer
  mercury, 75
  resistance, 77
time base, 41
tolerance, 11

torque measurement, 71
total error, 11
transducers, pressure, 142
transistor, 105
transmission, 147
triggered time base, 43
truth tables, 123

velocity measurement, 138
venturi meter, 92
vernier
  caliper, 62
  scale, 63
voltmeter, 35

wattmeter, 31
  corrections, 33
Wheatstone bridge, 10, 44

yard, 1

zero temperature, 4